THE LANCASHIRE AND CHESHIRE MINERS

THE LANCASHIRE AND CHESHIRE MINERS

Raymond Challinor

FRANK GRAHAM
6, Queen's Terrace, Newcastle upon Tyne 2

Published 1972
SBN 902833 545

Printed in Great Britain by
Northumberland Press Limited
Gateshead

CONTENTS

LIST OF ILLUSTRATIONS

CHAPTER ONE

The Beginnings of Trade Unionism among the Lancashire Miners

Evidence suggests that the Romans were the first to use coal in Lancashire. Traces of ash from coal fires have been discovered in Roman remains in Manchester. It also seems that coal from the Wigan area was transported for use in Warrington.[1] But this happened on an extremely small scale. The abundance of wood made coal an unnecessary substitute, particularly since it often required arduous winning.

The first written reference to coal came in 1246. Adam, son of Alexander, took legal action against Adam de Radcliffe for stealing four acres of pasture, where he dug for 'minera'.[2] Significantly, this is only ten years after the first recorded mention of coal mining in England.[3] So it seems that Lancashire was to the forefront in developing this new source of fuel. From the thirteenth to sixteenth century, there are many scattered references to coal and coalpits. Its importance, however, appears to have been negligible. Almost certainly production remained low: Professor Nef estimates each pit's total output would be worth between five and fifteen pounds.[4]

The low level of demand limited production. People preferred to burn wood. When this was not available, they chose turf or, as it was called in Lancashire, 'flat earth'. Only as a last resort would they turn to coal. The reason for their aversion was quite sound. Their homes usually had a fire in the centre of the room and a hole in the roof through which they hoped the smoke would percolate. Normally, this created an uncomfortable atmosphere; when coal was used it became virtually intolerable. For the coal mined in the Middle Ages was of very inferior quality. While it produced little heat, it did produce dense black clouds and acrid fumes. Hapless householders condemned themselves and their families to smoky suffocation.

Craftsmen augmented the meagre domestic requirements. When lime-burners came to build a castle, fortress or cathedral, they needed coal. But their demand was intermittent and localized, confined to the immediate vicinity and period of actual construction. A steadier demand came from the smiths. They discovered that coal provided a constant, hot flame, quite suitable for rough work. It was more economical than wood, but its greasy content might spoil the metal. So smiths tended to use charcoal for their finer pieces.

All told, coal consumption in Lancashire at the beginning of Queen Elizabeth the First's reign was probably around 10,000 tons a year. By 1700, it had increased fifteen-fold.[5] This rapid expansion can be partly attributed to a growing shortage of timber, which pushed up the price of wood and caused people to search for a substitute. Fortunately, this coincided with improving mining techniques. These could be applied with greater ease because of the enclosure movement in Tudor times that concentrated land-ownership into fewer hands and facilitated operations on a much larger scale.

There is plenty of evidence great wealth was amassed while the industry was still young. In 1538 John Leland, in the course of his travels, observed, 'Mr Bradshaw hath a Place called Haigh a myle from Wigan. He hath founde moche Canel like Se coal in his grounde very profitable to hym.'[6] Confirmation of this fact comes from the Haigh accounts: to start a pit only cost a few pounds and it paid for itself within a year.[7] Parliamentary commissioners visiting North Lancashire found a similar situation there. Since the reign of Elizabeth the First, small collieries had started within a dozen manors and the industry was in a healthy state. The commissioners reported that a Haslingden mine, with a rent of five shillings, made a profit of £60; one at Padiham, rent £2 13s. 4., made £40; another at Ightenhill, rent six shillings, made £20; and one at Trawden, with rent eleven shillings, had a profit of £60. As the value of mineral rights on crown land appreciated, the commissioners recommended increased rents for mining but not for agricultural land.[8]

Like the crown, landlords did not benefit initially from the industry's expansion. They received fixed annual payments, and had to wait till the leases expired before they could raise rents. But when the opportunity did arise, they exploited the market to the full. Where five to ten pounds had been regarded as a high valuation in 1575-1600, concessionaries were prepared to pay a hundred pounds by 1630.[9] The swift rise in rents aroused land speculation. Men like Robert Leigh made their fortunes in this way, which only served to push rents even higher.

But not everyone benefited from the new prosperity. Agricultural tenants frequently found their ancient rights had been abrogated. The enclosure of common land and the taking of pasture for mining operations endangered their livelihoods. Sometimes they reacted to this threat violently. Robert Bold, who worked coal in the waste at Sutton manor, had his fences destroyed sixteen times by people claiming it as common pasture. Eventually, a court sanctioned enclosure, but ordered Bold to see that there was no danger to wandering cattle.[10] At Colne manor in 1574, James Hartley and others entered the land of Anne Townley. They marched to her coalpit. They assaulted miners, broke baskets and winding-gear, threw timber and earth down the shaft. Anne Townley's servants remained at the pit-bottom, 'not beinge able to wynd themselves owt'. The intruders threatened to set fire to the pit with straw and gunpowder next time if she continued to make encroachments on the common land.[11] At Atherton the lord of the manor sued his tenants, who resisted his 'getting, bankeing and takeing of coales'. The tenants claimed that their right to mine coal was not excluded from the original deeds nor was the lord entitled to 'freedom of passage, egress or regress'. The lord, who was unable to produce the original deeds and claimed that he had been informed of his rights by elderly persons, won the case.[12]

Court decisions reinforced the prevailing trend for the ownership of land to become concentrated in fewer hands. Many of the dynasties of mine-owners—the great families who dominated the industry in the nineteenth and twentieth centuries—began their activities in this earlier period. Having acquired the land and capital during the reigns of the Tudors and Stuarts, they proceeded to make lots of money. Only the very affluent were able to afford the expensive winding-gear, pumping apparatus and exploratory excavations, which subsequently became necessary as the scale of operations grew.

The whole structure of the industry, and its *modus operandi*, was transformed as it expanded. In the thirteenth century, reports mention the 'taking' or 'digging'—not 'mining'—of coal, which suggests it came from outcroppings. A labourer could easily get coal, which was lying on or near the surface. Only when these supplies were exhausted did he resort to actual mining. Then he might create a drift mine, a roadway into the hillside, or a bellpit, so-called because its shape resembled a bell. Such workings were shallow. As miners were obliged to venture farther underground, the problems of drainage and roof-falls grew. It therefore became the practice to abandon a pit once a few score tons had been won and sink a fresh shaft. Some parts of Lancashire today are honeycombed with relics of this early mining activity.

In the same way as the surface coal was exhausted, so was that of the shallow seams. It became necessary to go deeper and deeper. In 1600, the depth varied from 24 to 60 feet. By 1700, mines of 120 feet were quite common. Along with greater depth went greater costs. Originally, only a few pounds had been needed to sink a shaft; by the mid-seventeenth century it probably cost between £60 and £70; while by the beginning of the eighteenth century the expense might range from £100 to £1,000.[13] No ordinary collier could afford such a sum. This task inevitably fell to a new and emerging class, the capitalists.

Certain problems seem to be endemic to coal mining. They have plagued the industry from its inception. In the early stages, when geological knowledge was very rudimentary, it was often a hit-and-miss affair whether the shaft ultimately struck the seam. Even if it did, difficulties were far from being over. With only one shaft, there could be no through current of air and dangerous gases tended to linger in the workings. So a collier, swathed in damp bandages and holding a long, lighted stick, was sent down to explode the gas before work began. It was a hazardous occupation which led to a high mobility of labour. Nevertheless, in spite of these precautions, explosions still occurred during working hours. In 1688, Tom Winstanley, the manager of the Haigh pits, reported to his master, Sir Roger Bradshaw, that 'the fiery damp went off twice, but did little hurt'.[14]

Another hazard, as burial records amply testify, was falls of roof. Again in 1688, Tom Winstanley complained that one of the pits under his supervision—Rough Hey pit—was suffering from frequent roof-falls: 'It falls over their heads almost every day; which causeth so much dressing, it eats away ye profitt.'

A further source of trouble in many pits was the presence of water. Sometimes owners sought to drain it by using pumps, operated by horses, or, where possible, they directed it into falling streams. Another method was to drive narrow tunnels from the workings into a nearby valley. These tunnels were generally referred to in Lancashire as 'soughs'. Most impressive was the Great Haigh Sough, completed in 1670. It ran from the brook at Bottling Wood to Park pit, a distance of 1,121 yards. Sir Roger Bradshaw, who claimed he built it for posterity, ordered that there should be a monthly inspection. His instruction was carried out until 1923, when subsidence made further inspection impractical. The closure of the irrigation complex based on the Great Haigh Sough eventually led, in 1932, to the flooding of most of the pits in Aspull and Westhoughton.[15]

But, besides technical things like ventilation and drainage, mining also had to content, from its early days, with human difficulties: one of the problems of mining is the miner. Perhaps because of the arduous and hazardous nature of his work, he has always been prone to absenteeism and insubordination. Some complaints about the behaviour of colliers, which were written almost three hundred years ago, have almost a contemporary flavour. For instance, in 1688 Tom Winstanley told his master that men were claiming wages for more work than they did: 'Whereas when three, fower or five Hollydayes happen in one weeke. And when ye workmen Idle three, fower or five daies in one weeke, it has been a sham practice of ye winders and workmen to come and worke one day; and to call yt one week's worke.' He advocated that Sir Roger should pay by the day, not by the week: 'For ye Labourer is worthy of his Hire; and ther's no Reason hee should Receive pay who does not work.'

Winstanley was also concerned about colliers' industrial conduct. He wrote to his employer complaining: 'I am sometimes tormented with the damnable workmen ... They are the arrantest knaves in nature. How you approve of the method I have taken with them, I pray you, Sir, advise, I have taken off all allowances of cannel.'[16]

Notwithstanding such reports, it would be wrong to imagine the main conflict in this early period was between master and servant. Most mines were small, employing a few men. The owner knew his men personally; often they were united by strong mutual ties of loyalty. Their main opponents, the people with whom they usually came into conflict, were the owners and colliers of neighbouring mines. The difficulty of moving coal any distance, the limited size of the market, threw each colliery into intense competition with others in the vicinity. A small fluctuation in the quality or price had a direct impact on nearby mines, and this could lead to strife. For example, in 1598 William Orrell advanced prices at his mine near Wigan. His neighbour, Katherine Sherrington, saw an opportunity to increase her trade at his expense. 'Having a coal pit in her said close, she opened up the same for the good of her poor neighbours and not for any profit to herself.' Enraged, Orrell retaliated by firing four arrows at her. Then she, it was alleged, told her workmen to tunnel under the highway, steal some of Orrell's coal, and smoke out his workmen with straw.[17]

This type of irregular warfare was quite common. Disputes did not always arise from trade rivalry; production itself provided a fertile source for contention. The epic quarrel between Bradshaw, of Haigh, and Gerrard, of Aspull, exemplifies this point. Gerrard claimed that water, coming from Haigh, was flooding his workings and damaging his sough. He approached Bradshaw to remedy it. But Bradshaw refused, arguing that, with the lie of the strata, there was a natural flow of water from Haigh to Aspull. This did not satisfy Gerrard. A fire mysteriously started at Bradshaw's Stone Lowe Pit. Whether Gerrard was responsible or not is uncertain. Nevertheless, it is clear he did send his men on night raids. They stole coal from the pillars, supporting the roof, so it became impossible for Bradshaw to get coal from some of his workings. On one of these secret nocturnal expeditions, Bradshaw's men discovered Gerrard's men in the act, and a fight ensued.[18]

For centuries, disputes of the Bradshaw-Gerrard kind occurred. An Act of Parliament in 1740 endeavoured to stop them. It declared that 'anyone maliciously damaging his neighbour's mine by water was subject to pay treble damages and full costs'.

The law, however, was not able to cope with another source of discord—the public's anger at the destruction to, and congestion of, roads caused by the constant carrying of coal. Horses, with panniers strapped to them, were the normal means of transport, and many journeys had to be made before any sizeable quantity had been moved, hence trouble for other road-users was extremely likely. In 1667, coal taken from Prescot to Liverpool for shipment to Ireland gave rise to complaints. As one writer observed, 'the streets are much decayed and abused by the frequent drivings of carts laden with coals'.[19] In the eighteenth century, when Arthur Young made his tour of England and Wales, he found the road between Preston and Wigan in a very bad condition:

> 'I know not, in the whole range of language, terms sufficiently expressive to describe this infernal road. To look over a map and perceive that it is a principal one not only to some towns but even whole counties, one would naturally conclude it to be at least decent; but let me most seriously caution all travellers who may accidentally purpose to travel this terrible country to avoid it as they would the devil, for a thousand to one but they break their necks or limbs by overthrows or breakings down. They will meet ruts which I actually measured four feet deep and floating with mud only from a wet summer; what, therefore, must it be after winter? The only mending it receives is the tumbling in of some loose stones, which serve no other purpose but jolting a carriage in the most intolerable manner. These are not merely opinions but facts, for I actually passed three carts broken down in these eighteen miles of execrable memory.'[20]

Heavily-laden coal carts not only injured Lancashire's bad roads even more, but were exasperating for the coalowners as well as the public. The numerous breakdowns of coal carts and the long time taken by journeys meant that transport costs were high. The whole expansion of the industry depended upon the provision of a better form of transport.

The search resulted in the development of waterway traffic. Between 1719 and 1727, the river Douglas was straightened and deepened. It became a commercial artery, providing the means for the rapid expansion of coal mining in the Wigan area. In 1775, improvements to the Sankey brook had a similar effect on St Helens, making access to Liverpool much easier.

Then the great age of canal-building began. In 1759, the Duke of Bridgewater complained to Parliament that it was impossible for him to exploit his mines efficiently at Worsley. To carry coal from there to Manchester, less than seven miles, cost nine to ten shillings a ton. Parliament gave assent to the Bridgewater Canal Act. Originally the duke intended to use a little stream called the Worsley brook, which flows into the Irwell near its confluence with the Mersey, But he abandoned this idea on the advice of James Brindley, a Staffordshire engineer, who had recently drained Fletcher's Clifton mines by installing a water-wheel. Brindley was a man with practical genius. Almost illiterate and without knowledge of the scientific movements of his time, he still possessed high intelligence combined with great powers of concentration. He seldom made drawings or plans, but relied upon his memory. When problems arose, he spent days in bed cogitating over the solution. Despite these somewhat strange procedures, Brindley built a new, revolutionary canal.[21] It began as underground galleries in the depths of the coal strata and reached Manchester by crossing the Irwell by a forty-foot aqueduct. Completed in two years, the general public regarded the Worsley canal as the eighth wonder of the world.[22]

Encouraged by his success, the Duke of Bridgewater became the leading figure in promoting canal construction. In less than thirty years, the contours of England were furrowed by navigable waterways. The Leeds-Liverpool canal, crossing the Pennines, permitted Yorkshire coal to compete in the Lancashire market. The Grand Trunk canal linked the Irish Sea and gave the North Sea via the River Trent. This gave Staffordshire coal access to the markets of Manchester and Liverpool.[23] Lancashire became the focal point of the canal network, with small ones like the Bury canal and the Bolton canal as well as the long one extending through Preston and Lancaster to Kendal.

From 1750 to 1800, the development of the canal system was closely associated with the growth of the coal industry. The reason for this is to be found in the nature of the commodity; its size and weight was largely relative to its value. Consequently, the cost of carrying coal added greatly to the selling price. When the opportunity arose to make economies, coalowners naturally seized it. For the Duke of Bridgewater, the cost of transporting coal to Manchester dropped from twelve shillings to six shillings a ton.[24] The Trent-Mersey canal resulted in even greater savings—to about 12½ per cent of its previous price.[25] As a result, coalowners did their utmost to utilize canals. In some places, such as Standish and Worsley, the waterway actually ran into the pit.

Canals acted as an immense stimulus to production. Speculators were attracted to the industry by the prospect of quick profits. In a spirit of euphoria, they opened many new mines in Lancashire. But the old-established coalowners, accustomed to a steady income, feared that a big increase in output might ultimately cause a crisis of over-production. Men like Lord Balcarres bought out smaller owners merely to restrict competition. He acquired mineral rights, sometimes concessions of dubious value, to prevent others from exploiting them.[26] Where this proved to be impossible, he sought to bankrupt his smaller rivals by having a price-war. Besides possessing the advantages that accrue from economies of scale and employing more technically proficient staff, Lord Balcarres' political influence placed him in a privileged position *vis-à-vis* his competitors. On his suggestion, the Lancaster Canal Act 'exempted him from certain duties on coal won from his own estate, within the township of Haigh and adjacent to the canal; this was worth threepence a ton'. As his adversaries were compelled to sell out, Lord Balcarres' empire continued to expand. The same trend was manifest throughout Lancashire: large scale production and a smaller number of producers. By the beginning of the nineteenth century, the industry started to acquire a distinctly modern appearance.

These changes coincided with the quickening tempo of the entire economy. The developing of textiles and the iron industry greatly increased the amount of coal required. Demand and availability rose rapidly as the potentialities of the steam-engine were realized. The whole character of Lancashire underwent a transformation.

Trade unionism is usually a concomitant of the development of an industrialized society. Yet, in the mines of Lancashire it was slow to appear. One historian expresses surprise at its delay, as if it should have been a reflex action, a response to the accumulation of capital.[27] But there is nothing automatic or inevitable about it. Certain necessary conditions must be present. Of these the most important is an economic cleavage between capital and labour; ownership of the means of production being vested in one section of society while another section is compelled to sell its labour-power in order to live. In these circumstances, potential conflict is inherent. Either side—capitalists or workers—can seek to improve their relative position at the expense of the other. But this does not necessarily happen. Existing arrangements may be hallowed by tradition so neither side disrupts prevailing practice, having been accepted in the past it is accepted in the present.

Where opportunity for class mobility exists, there is less likelihood of class conflict emerging. The more intelligent and enterprising workers—the men who, in a different situation, would provide the vital cadres that formed the union leadership—merely better themselves. They solve their problems in a personal rather than collective fashion. Instead of formulating and articulating the grievances of their fellow workers, they simply strive to improve their own position, rising from their class, not with it. However, where such opportunities for self-advancement do not exist, a different attitude may easily arise. Realizing that they are condemned for life, whatever their efforts, to be members of the working class, they begin to identify their own self interests with those of the class: to improve one improves the other.

A further condition necessary for trade unionism is the existence of occupational stability. If men frequently move from one job to another, they are unlikely to possess the appropriate outlook. The formation of a union requires men to identify with each other, have confidence and trust, and be concerned about winning improvements. All these attributes are nurtured by working together over a long time; it is improbable that they will arise in men whose contact with each other is merely transitory.

But in the early days of coal mining this is what usually did happen. Frequently colliers combined farming and other work with that of being a pitman. So when mining compared unfavourably with other occupations, then the tendency was to leave it. No attempt would be made to exert pressure for better pay and conditions; the easiest way, as well as the course of least resistance, was to find work elsewhere. In Lancashire, with its highly variegated economy, this was easier to do than in most other coalfields.

A fourth, and final, condition is the development of an awareness, a consciousness that gives the union stability and durability. In the eighteenth century it is possible to cite examples of violent struggles by workers, but they tended to be over a particular grievance. Once it was settled, workers would subside into apathy, leaving no lasting organization to commemorate it. Only with the emergence of the idea of social class, a concept that appears to have originated in the French

Revolution, can there arise the consciously-felt need for permanent organization on a class basis.

With the Lancashire miners, it is easy to see the forces that retarded the growth of trade unionism. In the industry's formative period, the prevalence of small-unit production, the paternalistic attitude of the master to his men, and the rivalry between one mine and another, all helped to prevent workers from acquiring a sense of unity.

To make matters even more difficult, employers introduced a system of fines. While this was intended to be a means of enforcing industrial discipline, it was also a useful device for dealing with any incipient trade unionist. Any man questioning the wisdom of any move made by his employer was liable to find himself arbitrarily fined, a chastening experience as the penalty imposed customarily constituted a large proportion of his income. When the average daily wage at Haigh was two shillings for miners and a shilling for labourers, a man would lose tenpence for abusing a customer, sixpence for removing coal in a basket, sixpence for taking away too much of the supporting pillars, and sixpence for making false returns. Similarly at Worsley: when the Duke of Bridgewater paid colliers eight shillings a week for six days of between twelve and fifteen hours each, he fined men two shillings and sixpence for being late for work on Monday morning. Money collected in fines went to build cottages, a number of which were known as 'Half Crown Row'.[28]

Another sanction coalowners employed, which could be used against men trying to form a union, was to disallow some of their wages on the pretext that the coal they had hewn contained dirt or was of poor quality. At Haigh, colliers complained that they were only paid twopence for fourpence or fivepence worth of coal, 'which was a horrid cheat and shame'.

Any miner showing undue concern over the welfare of his fellow workers was almost certain to be victimized. Probably, his master would dismiss him. But even if the master merely demoted him to labourer or drawer, then his wage would be halved, a harsh penalty for insubordination. By contrast, the ambitious men, prepared to forget about bettering others and concentrate on bettering himself, had a number of opportunities. He might become a chartermaster or 'butty'. Then he would be in charge of actual production. He would hire workers, supervise them, and pay them. In return, he would have a contract with the coalowner, stipulating the price he would receive for a given quantity of coal. As a result, the coalowner's concern became primarily commercial: to find the capital and market the product. He did, however, in most instances determine, in broad outline, the methods of work. Where production was on a large scale, the owner usually appointed an official at each pit—he was called 'the auditor' at Haigh—to maintain general surveillance. Nevertheless, there can be no doubt that butties enjoyed considerable power and, compared to ordinary miners, received a high income[29]

Throughout the history of coal mining, the butty system remained an important influence militating against trade unionism. It created divisions among the workers. An intermediary—the butty—became the focus for protests over pay and conditions that rightfully should have been directed at the owners.

In mines where the butty system did not operate, there were still openings for the man who aspired to join the aristocracy of labour. If he worked in a small pit, employing a dozen or less, he might aim ultimately to become the 'banksman' or manager. In larger enterprises the opportunities were commensurately greater. A 'viewer' would be appointed as general supervisor of underground work. The 'overman' would be responsible for all operations not connected with actual mining—known as 'dead work'—such as sinking shafts and attending to soughs. As the scale of mining grew, the echelons of management likewise became more extensive. A greater differentiation within the working class, better prospects of promotion helped to keep ambitious workers enthusiastic and obedient.

Another method coalowners applied to retain a pliant proletariat was the bond. This was a contractual arrangement whereby men agreed to work for a certain employer for a given period. In 1636-7, Wigan colliers bound themselves for twelve months to dig at the Bishop of Chester's pits at Farnworth. They received a signing-on fee of half a crown.[30] At another mine, in 1676, a collier was called upon to 'well and truly serve a year in the coal mines of Hulton and Denton, according to the customs and orders they used, or to forfeit forty pounds sterling'.[31] In 1750, two shillings was paid 'for hiring Henry Bramwell for two years as a drawer at Haigh' whilst, in 1787, at Gilbert Close pit, Haigh, six miners were bound and received half a guinea each.[32]

Professor T. S. Ashton and J. Sykes suggest that binding was mutually advantageous in a period when transport was primitive and conditions in the industry rarely changed. It assisted employers, assuring them of a regular supply of labour; it guaranteed the miners a certain wage and working arrangements. Ashton and Sykes argue that the system only ended when miners 'ultimately came to rate freedom to work or not to work, freedom to move and freedom to combine above the mere maintenance of safe and customary standards'.[33]

This view, in my opinion, is open to question. It appears that Ashton and Sykes relate the end of binding to the industry's development. Yet, the most advanced coalfield, with the most sophisticated production methods and the largest output, was in the North-East, where owners tenaciously clung to the practice until the second half of the nineteenth century. It is hard to believe the bond gave workers any tangible benefits. It is impossible to cite many instances where miners demanded the introduction of the bond where it did not exist. There were, on the other hand, numerous occasions when men called for the abolition of the bond. The strikes of 1765, 1810, 1831-2, 1844 and 1858 in the North-East coalfields were all against binding.[34]

Likewise it is difficult to accept the assertion of Ashton and Sykes that the bond guaranteed certain safety standards. For men were compelled to work, on penalty of being fined or imprisoned, whatever the conditions happened to be. Inadequate ventilation, the presence of gas, the danger of roof-fall were ignored when an owner had a profitable order to fulfil.[35]

The masters found binding to be a very valuable weapon for enforcing discipline. They were reluctant to discard it. The reason why the bond ended in Lancashire before the North-East can probably be attributed to the greater problem of discovering defaulters. There are many instances of Lancashire coalowners recapturing absconding colliers. But in this county the task was immeasurably harder than in Northumberland and Durham, where coalowners were united in a mighty association—the Vend—which regulated prices and output. Any miner there who left work without authorization would find it impossible to gain employment at another colliery. However, in Lancashire, where employers were in competition with each other, a miner would encounter no such obstacle. Indeed, he might even receive a friendly welcome from a rival coalowner. Furthermore, in the North-East pitmen lived in closely-knit communities whereas Lancashire was much more diversified. Hence, men 'on the run' had a better chance to get employment in another occupation. It may also be that the state of the labour market, the influx of Irish immigrants and agricultural workers, made Lancashire coalowners feel they had not the same need to go to such lengths to secure workers.

In Lancashire binding was never universally adopted and, even where it was applied, the practice seems to have died out in the first quarter of the nineteenth century. Nevertheless, so long as it did exist, there can be no doubt that the miners regarded it with intense hostility. Whatever the injustice, however great the exploitation, miners working under the bond had to continue their labour. It denied them one of the main things that distinguishes free men from slaves—the right to withdraw their labour.[36]

The first large disturbance involving colliers appears to have occurred in June 1757. For many years, wages had been rigidly fixed. When suddenly the price of corn, a basic commodity, rose, it was tantamount to a pay reduction. In most coalfields hungry colliers reacted with instinctive violence. Their contempt for danger, an attitude engendered by the hazardous nature of their own occupation, did not lead them to attach an exaggerated value to the lives of people who crossed them. At the same time, their work toughened them so they did not flinch from physical encounters. Angry and starving, miners were a formidable foe for the forces of law and order. In Bristol, there was what *The Gentleman's Magazine* described as 'an insurrection of the colliers', with the city being occupied for two days.

By comparison events in Lancashire were less serious. Colliers marched from Clifton in the direction of Manchester. But the authorities, assisted by some alarmed citizens, barred their way at Salford Bridge. Some of the demonstrators, however, appear to have found other ways of reaching Manchester. Their presence did not please local inhabitants. The colliers were chased up Oldham road, and two of them, who were caught at Shudehill, were 'handsomely Drubbed and turned out'.[37]

Significantly, in these early struggles, miners did not attribute their plight to the coalowners. It was rather some ill-defined enemy, perhaps society in general, which they blamed for their distress. As they had no notion of class consciousness, they strived for industrial unity, a united front of masters and men. To some extent, this aim was achieved. Some coalowners gave their blessings to miners' protests. Many of the marches were led by overmen and butties, appointed as 'captains' in charge of the workers in the demonstration. It is not surprising, therefore, that when there was a further food riot in 1779, weavers smashed spinning frames at Pemberton but miners did nothing to damage colliery property.[38]

Thirteen years later, when there were the first serious overt acts of class conflict, the situation radically altered. Coalowners were worried both for themselves and their property. Trouble began when the men who drove the coal barges demanded a shilling a ton per trip. Henry Blundell, owner of Orrell collieries, wrote to the Home Secretary for advice. The reply came 'that you should pursue every legal and constitutional means of suppressing such combinations and bringing the leaders to punishment'.[39] The Home Secretary ordered a troop of soldiers, stationed at Manchester, to be ready to move to Wigan if Blundell deemed it advisable.

Discontent in the district apparently continued. After a further four months, Henry Blundell again reported that there was industrial unrest:

> 'The agents from the extensive collieries in the neighbourhood of Wigan (and on which the town depends for its fuel) came this evening with the news of the colliers having left their work and collected in a riotous manner to the number of near five hundred; and had been with them thus assembled to demand an extravagant advance of wages. They have given them only till tomorrow at three o'clock to consider it, and if their demand is not complied with, they threaten to destroy the Works by pulling up the engines, throwing down the wheels and filling up the pits. The consequence to coal proprietors would be very serious and to the country in general if this combination is not immediately suppressed ...'[40]

At that time, according to Blundell, a miner and his drawer could earn them from five to eight shillings a day. They wanted a further two shillings. Their threat to destroy machinery was a common stratagem in an age when workers' organizations were weak. With scant financial resources, the men could not hope to sustain a prolonged strike. An attempt to intimidate provided their best, even if slender, chance of success. 'Collective bargaining by riot', to use Dr Hobsbawn's phrase, was the most effective means of pursuing a wage claim.[41]

In this instance it failed. The arrival of the military deterred strikers from attempting to destroy colliery machinery. After being out a week, they returned to work.

Soon miners learnt a lesson from the defeat. In October 1794, the first association of miners—the Brotherly Union Society—was formed. It met at the Ben Jonson Inn, Pemberton. The Society's rules were amended in October 1804, and again in December 1830, an indication that the Society had a long and active life. After a while, other colliers sought to copy the idea. By the end of the eighteenth century at least ten similar organizations existed in Central Lancashire and, a few years later, the number rose to twenty-one.[42] They functioned as friendly societies in an effort to conceal their trade union activities from the authorities. Some even conducted business in their wives' names as a precaution against prosecution under the Combination Act.[45]

Such circumspection, however, did not save them. In 1818, when a dispute involving thousands of miners swept through the entire Lancashire coalfield, the authorities resorted to repression. They first realized that something was amiss when 700 colliers attended a meeting at Kersal Moor. On the following day, 19 August, 1818, a printed handbill appeared. It stated that workers from many different industries had resolved to form a unified organization, to be known as the Union of all Trades.[44] Events then moved rapidly. First the weavers stopped work, followed by the miners. Within a short time, Lancashire's industrial life came to a halt.

As so often happens, the employers did not believe the unrest originated from bad conditions. In their opinion, the workers had nothing to complain about and, consequently, the employers had nothing to rectify. With this comforting thought in mind, the coalowners could devote themselves to denouncing even more vigorously, the interlopers who they thought were responsible for the trouble.

Outsiders, it must be admitted, did play a prominent part. At the colliers' meeting on Kersal Moor, two Methodist preachers, Benjamin Newton and Wiseman, were among the speakers. It was after one of these ministers had spoken, quoting heavily from the Scriptures and likening the colliers to the children of Israel in bondage, that the decision to strike was taken.[45] At Oldham, Captain William Chippindale, of the Militia, reported to the Home Secretary that he had discovered 'that the notorious Dr Healey has been tampering with the workmen of the Alkrington colliery, a concern of which I have the management ...'[46] He said that Colonel Fletcher, a Bolton coalowner, had important information that 'establishes the fact of the interference of the Jacobins'. Another coalmaster, James Norris, attributed his misfortunes to 'that pest of our happiness and freedom—the itinerant radical reformer'.[47]

Notwithstanding such remarks, there can be little doubt that colliers strongly and sincerely believed that their grievances were justified. It was upon their own efforts that they principally relied. A Manchester conference, attended by 120 delegates elected by colliers, formulated their demands, which it was agreed to put to the coalowners.

When a deputation interviewed Colonel Fletcher, he accused them of belonging to an illegal combination and had four of them arrested. They were prosecuted for breach of contract, and sentenced to three months' imprisonment each. Their fellow-workers, angered by Fletcher's conduct, vowed 'they were going to bring him down on his bare knees for this'.[48]

But Fletcher saw it his duty to defend the prerogatives of property.

'The danger of a committee dictating what wages must be paid' would be, in his opinion, 'worse than universal suffrage ... It would introduce a mob oligarchy, bearing down all the better orders of society, and would quickly be succeeded by universal anarchy.'[49] He was annoyed that some owners, out of what he described as fear and criminal apathy, had capitulated to the men's demands. Yet, despite these strong words, he also succumbed to pressure. Not only did he concede wage increases but petitioned the Home Secretary for the release of the imprisoned strikers.[50] If he adopted a conciliatory attitude, Fletcher thought that his colliers would renounce trade unionism, a belief which was as optimistic as it was wrong.

By November 1818, the turmoil in Lancashire had subsided, with the workers at many pits winning concessions. Their success, an illustration of their potential power, served as a spur to others. St Helens miners, remembering a wage cut received during a trade depression three years earlier, approached the owners for an advance. The coalmasters replied, that, as they now employed a quarter more men, they could not afford to grant it. This answer did not satisfy the miners. They decided, therefore, to strike at successive collieries; the men still at work contributing to the strike pay of the others. News of this plan reached the employers, who resolved to forestall the miners by locking out all men known to be unionists. Quickly the dispute became general throughout the St Helens area.

On 17 February, 1819, after the struggle had been on for six or seven weeks, Michael Hughes and Colonel George Williams, two local coalowners who were also magistrates, wrote to Lord Derby, lord lieutenant of the county: 'all calculations and reasonings upon the subject we have found unavailing to induce them to return quietly to their employment'.[51] Lord Derby dispatched their letter to Viscount Sidmouth, the Home Secretary. He added, 'I flatter myself the measures which have been taken by the magistrates will preserve the future peace of the country and put a stop to these illegal combinations ...'

The main action appears to have been to arrest John Johnson, alias Cooke, the strike leader. He was arraigned before Hughes and Williams, charged with administering illegal oaths. A crowd of colliers surrounded The Raven Inn, St Helens, where the proceedings were taking place, and there was fear that an attempt to rescue Johnson might be made. This was foiled: Hughes went to the front of the inn to address the multitude while a policeman drove the prisoner away in a chaise from the back. As soon as the miners became aware of this trick, the *Liverpool Mercury* says:

'The mob became very clamorous, and followed the chaise, but without being able to accomplish their object. The men then dispersed; but we understand that in the evening they visited the pits in large bodies, with fire-arms and put out the fires and discharged the engineers at the different works. Alarmed at these proceedings, the constituted authorities thought it necessary to call in the aid of the military; and on Saturday evening, a company of the 52nd regiment and the Liverpool Light Horse received orders to march for St Helens. A party of infantry also marched from Warrington for the same purpose.'[52]

The military remained in St Helens to protect property, and more union leaders were arrested. At this point, strikers appear to have become demoralized; the drift back to work began. The magistrates sought to stimulate it by assuring all strikebreakers of full support and protection. They also attempted to drive a wedge between those still on strike and the rest of the workers. Placards posted in the town pointed out that the actions of well-paid colliers were putting less well-paid auxiliary workers, like waggoners and drawers, out of work. The authorities, moreover, threatened to confiscate the funds of friendly societies assisting the strikers. Since many societies were not confined to miners, this announcement must have caused apprehension among the rest of the membership. Men from other trades, fearful lest their precious boxes were seized, doubtless urged colliers to be more cautious. By 2 March, Hughes could report to the Earl of Derby that the colliers have 'returned to their work as usual, without the smallest injury having been done to the person or property of any individual'.[53]

From the men's standpoint, the nine-week strike was a failure. They returned without having secured an increase. Yet, as Dr J. R. Harris has stated, to sustain a struggle of such magnitude and duration, so soon after a period of trade depression, shows a remarkable degree of organization.[54] The use of what has become known as the Lancashire strike tactic, a succession of strikes supported by those still at work, indicates it was not an instinctive revolt but planned with considerable sophistication.[55]

In the next few years, the coal industry in Lancashire remained relatively peaceful. Colliers gave financial assistance to the cotton spinners' strike in 1823. Three years later, miners themselves chafed when owners introduced wage cuts. Strikes and riots proved to be of no avail. Their efforts were far too disorganized and unco-ordinated.

This was not true, however, in March 1830, when South Lancashire colliers met at Bolton to form the Friendly Society of Coal Mining. The union structure was based on local lodges, each with a committee of seven elected from among its members. Every lodge sent three delegates to the quarterly meetings, where one of the most important items for discussion was the report of a sub-committee appointed to investigate pay at the various pits. The union's efforts seem to have been concentrated on the wages issue; no provision was made for sickness or accident benefits.

Under the general direction of the Authority lodge—the leadership with its headquarters in Bolton—the union's influence spread. William Twiss, one of its best organizers, went to South Wales to recruit members.[56] When, in April 1831, another conference was held at Bolton, delegates from Lancashire, Yorkshire, Staffordshire, North and South Wales attended. Subsequently, it established links with the Derbyshire miners.[57] But no contact appears to have been made with the colliers of the North-East, where Tommy Hepburn had created a powerful organization.[58] If such links had been established, then the miners would have built their first national union.

As it was, the Lancashire lodges, the backbone of the union, seem more concerned with achieving unity with workers in other industries rather than extending the solidarity among workers in their own. This was probably because of the need to relate themselves to the diversified nature of the Lancashire economy. It is also the reason why they gave enthusiastic support to the National Association for the Protection of Labour, a body led by John Doherty, which aimed at securing a greater degree of harmony and joint action among the workers of all trades. Operatives from twenty trades joined, including the Blackrod Coalmen.[59] Miners' representation increased in 1831, when the Bolton conference decided to affiliate the entire membership.[60]

The increased power and cohesion of the Lancashire colliers may have tempted them to be more aggressive. Or maybe the masters, alarmed by the union's growth, decided it was time to reassert their authority. Whatever the explanation, in the spring of 1831 the coalfield became engulfed by conflict. The same issue of *The Voice of the People* that carried the Bolton conference's appeal to all other miners —'We exhort colliers to continue cemented in one inseparable bond of union, doubting not that it will prove their strength and protection'— also mentioned that masters in Westhoughton and Standish refused to permit their men to join the union.[61] Similarly, Ashton-under-Lyne coalowners issued an ultimatum: leave the union or lose your jobs.

The men were equally determined not to leave the union. In the Bolton-Farnworth-Westhoughton area, they struck to defend their right to organize and also to demand an increase of one-third in their wages. One master, W. Hulton, published a handbill explaining his case:

'But while I acknowledge the divine authority of that beautiful instruction, "Let not thy left hand know what thy right hand doeth" (Matt: Vi.3), I may in self-defence assert that I have amply rewarded you for your labour; that I have never been so happy as in relieving yourselves and your families, under the pressure of sickness and distress; that I have to the utmost of my power extended the blessings of education to your children. I have a right then to ask: what has been your conduct towards me? You have wantonly injured me to the full limits of your ability, in my purse, and you have much farther wounded my feelings.'[62]

If many masters' feelings were the same as those of William Hulton, then no doubt many miners had feelings similar to those of the men who worked for Andrew Knowles & Son, of Pendlebury, who thought they were too poorly paid and too heavily fined. They could never earn more than fifteen shillings a week. Yet, they might be fined as much as seven shillings in a fortnight for failure to remove pit props after finishing work. If they were absent for a day, they lost half-a-crown. Whenever roof-falls occurred, miners were compelled, to pain of discharge, to clear the underground roadway for no payment.

The Pendlebury miners remembered what they regarded as an injustice which had happened eighteen months previously. A large quantity of gravel and water suddenly burst in upon the workings. The survivors rushed out and called on neighbouring pits for assistance. Together they cleared 70 yards of roadway which, by the force of water, was rapidly filled again. Driven back, the rescuers adopted a new tactic and cut through six yards of coal. There they discovered the bodies of an old man and his two daughters. After cutting another eleven yards, they discovered a further corpse. It was a wet and dangerous task. 'The men were to work night and day until they found the bodies. But when all the work was finished and the men came to be paid, they received only 1s. 3d. per day.'[63]

Like the Pendlebury colliers, men at many collieries had similar grievances that rankled in their memories. They saw the existence of a union as a means of getting retribution.

In most places the strike began in May 1831. Some coalowners made small concessions, others did not. A patchy, inconclusive outcome might have been the result had not the masters of Rochdale and

Oldham taken such a strong line. Instead of being content to maintain the *status quo*, they resolved to impose a wage cut. Moreover, where miners were not prepared to concede, strikebreakers were imported from Shropshire. 'Crowds of people flocked on Sunday to see the new colony of "baa-lambs", as the Rochdale folks call them,' reported *The Voice of the People*.[64] Escorted by the military, the strikebreakers were billeted in a decaying mansion called Stockfield. At Oldham, coalowners did not go to the trouble and expense of importing labour from Shropshire: they simply persuaded local spinners to go down the mines. In this way, they were able not only to maintain production but also do double shifts, a practice that Lancashire colliers spurned because of the health hazard. While collieries maintained output, the military stood by to protect them from attack. At night-time soldiers were stationed on Oldham church steeple and mine managements were told to warn them of any approaching hostile forces.

Coalowners openly avowed that they intended to smash the colliers' union. This they did. It could not cope with the situation. With miners in Lancashire and elsewhere preoccupied with their own local struggles, it proved incapable of providing a centralized strategy or giving a firm leadership. Also, the ease with which the coalowners obtained alternative labour and maintained production showed that the struggle was futile.

In August 1831, the Friendly Society of Coal Mining was no more.

Table 1 *Bridgewater Collieries Ltd* (1795-1810)

	Coal sold (tons)	Revenue	Expenses	Profit
1795	110,224	£23,681	£20,917	£2,764
1796	110,809	£25,783	£22,349	£3,434
1797	119,056	£28,307	£24,125	£4,182
1798	116,835	£28,008	£25,602	£2,406
1799	127,449	£30,867	£25,222	£5,645
1800	138,695	£33,978	£27,955	£5,023
1801	144,044	£36,568	£28,758	£7,810
1802	140,695	£44,742	£29,396	£15,346
1803	134,035	£55,477	£31,177	£24,300
1804	133,811	£54,848	£32,909	£21,939
1805	127,195	£52,714	£32,940	£19,774
1806	114,566	£44,764	£31,327	£13,437
1807	109,071	£39,629	£28,384	£11,245
1808	99,854	£35,083	£26,316	£8,767
1809	99,034	£32,971	£25,716	£7,255
1810	110,690	£44,067	£27,315	£16,752

Source: Company's book, Lancashire Public Records Office

Table 2 *Tonnage of Coal Entering Liverpool*

1781	34,401	1787	98,246
1782	39,326	1788	109,209
1783	57,376	1789	124,874
1784	70,555	1790	137,790
1785	85,138	1791	145,616

Source: Holt-Gregson Papers 10/215, Liverpool Records Office

CHAPTER TWO

The Miners' Association

The Miners' Association of Great Britain and Ireland, formally established at Wakefield on 7 November, 1842, constitutes an important landmark in the history of British trade unionism. Its size, structure and intention made it, in many ways, the prototype of the modern trade union, an advance on previous combinations of workmen.[1]

The Miners' Association had a longer continuous existence than any previous organization of working men in the coal industry. It lasted for at least seven years. Even when its enemies had smashed its national structure, when all sense of internal cohesion had vanished, remnants scattered in the various coalfields continued to use the name, 'The Miners' Association', and still regarded themselves as members. When the Leeds conference of 1863 was held, delegates did not consider they were forming a new union, merely revitalizing the old. (See Chapter four.) So, in an important sense, the Miners' Association never died—it left an indelible impression on all future combinations of colliers.

It could also claim to be national in more than name. At its zenith, the Association employed fifty-four full-time organizers, known as lecturers, and it possessed branches in all the major coalfields.[2] What is more, it endeavoured to formulate and implement a common policy, one which would be adhered to by all its constituent districts and lodges. In practice, this proved to be extremely difficult. Nonetheless, it did achieve some success in this direction, much more than any of its predecessors.

As a result, the Miners' Association had a greater impact on the industrial scene. It is impossible to calculate exactly what influence the Miners' Association had: to what extent improvements in pay and conditions can be attributed to its activities rather than to the workings of market forces. Yet, it can be confidently said that miners during the eighteen-fifties looked back nostalgically to the period when the Miners' Association was strong. For them, it was a golden age.

The union also tried to influence the legislature, endeavouring to persuade Parliament to pass measures which would improve the safety and health of pitmen. The passing of Lord Ashley's 1842 Bill, prohibiting the employment of women and small children underground, demonstrated that legislation could have tremendous impact, for good or ill, on the coal industry.[3] Consequently, colliers organized campaigns, petitioned Parliament, and briefed friendly M.P.s. The Miners' Association indulged in political activity to a degree hitherto unknown among British trade unions. Two of its leaders—W. P. Roberts and William Dixon—even stood as parliamentary candidates at the 1847 general election. However, in spite of the unprecedented concern shown, the political harvest reaped was negligible. It played an important part in mobilizing public opinion against the Masters' and Servants' Act 1844, an anti-union bill, which failed to secure a passage through Parliament.[4] It can also claim some credit for the Mines Act 1850, a measure that marginally improved safety and ventilation.

The Association was interested in another aspect of the law. It sought to protect miners from legal persecution by either the employers or the state. It engaged the services of a solicitor, William Prowting Roberts, a prominent Chartist who came from Bath.[5] A flamboyant and fiery personality, he became a popular leader of the union as well as its legal adviser. At one public meeting, William Dixon explained, 'If it were not for the legal talents of Mr Roberts, there was not a prison in Lancashire but would be full of colliers. If a warrant was taken out against a man, he said, "Oh, never mind, Uncle Bobby will defend me", and even if sent to prison, "Uncle Bobby will get me out again".'[6] Roberts saw his task as that of fighting an unremitting war against the 'King Coal tyrants': 'We resisted every individual act of oppression, even in cases where we were sure of losing.'[7]

The protection, both legal and industrial, which the Miners' Association afforded its members made it a source of wonderment and jealousy to contemporary trade unions. At its height, its membership reached almost 100,000.[8] Truly, the *Northern Star* was correct when in 1849, it described the Miner's Association as 'the most powerful Labour Union ever known in this country'.[9]

A complex interaction of forces led to the creation of the Association. This resulted in miners widening their mental horizons, jolting them out of their purely local preoccupations, and made them think of themselves as a class—a distinct and separate industrial entity with interests of its own.

The development of the Industrial Revolution, the growing reliance on coal and the steam-engine, gave the mining industry a crucial importance. The national well-being depended upon what it produced. As tables 2, 3 and 4 show, output continually expanded to meet increases in demand. And higher output meant more mines and more miners. As the labour force enlarged, it grew increasingly self-conscious and aware of its own value. 'We are the bees that make the honey,' declared one pitman.[10] Colliers compared their own impoverished state with the prosperity of the coalowners and industrialists; gentlemen whose wealth, directly or indirectly, arose from the miners' exertions.

The collier's sense of togetherness, his feeling for fellow colliers in other parts of the country, was fostered through improved communications. People travelled farther and more frequently. Colliers had a greater chance of getting to know other areas. Better roads and the advent of railways had a further effect. It facilitated the transportation of big quantities of coal. Coal entering Manchester by all conveyances rose from 737,008 tons in 1834 to 1,034,090 tons in 1840.[11] Developments like this led to the growth of a national market for coal. What happened in one district influenced the others. As a result, a miner could no longer afford to remain apathetic about miners elsewhere since their fates were intertwined.

The publication of Lord Ashley's report in 1842 merely served to confirm this point. Its nation-wide surveys showed that, while there were regional variations, poor conditions appeared to prevail everywhere. The factual data contained in the report possessed a common denominator in the colliers' minds: it corroborated their view that they had justifiable grievances. Public opinion was deeply shocked by stories of women, chained and belted like animals, dragging large loads from the depths.[12] It was also disturbed to read that little children, often only four or five years of age, toiled underground until they sometimes dropped asleep.[13] This public indignation, the extensively-held view that something was wrong with the coal industry, simply endorsed what miners had thought for a long time. Unintentionally, Lord Ashley's report, and the reaction of sections of the middle and upper classes, provided an added stimulus, siding the creation of trade unionism in the pits.

Further impetus came to the colliers from their experience during the 1841-2 slump, 'almost certainly the worst of the century'.[14] Hunger and unemployment were widespread among the working population. 'So deep was the distress, so universal the hunger,' said William Beesley, describing the conditions in North Lancashire, 'that they had to eat the dead carcasses of cows dug up after they were buried.'[15] In some parts of the North-East, families survived simply on a diet of salted potatoes.

Miners learnt a negative lesson in 1841: with a scanty, improvised organization, under conditions of labour surplus, it was impossible to prevent wage cuts. Local *ad hoc* bodies proved too weak. They resisted reductions—in January-February 1841 winding ropes were cut and the Earl of Balcarres' manager, Maskell Pease, twice had shots fired through his windows—but it was all to no purpose. The wage reductions relentlessly continued.[16]

By 1842, a wave of unrest spread throughout the entire country, apart from the Home Counties. Sporadic strikes started in the Midlands in May. By August they had gathered momentum, encompassing Staffordshire, Lancashire, Yorkshire and even Scotland.[17]

This conflict had considerable significance for mining communities. It gave them the practical knowledge of acting together, a sense of solidarity and unity of purpose that endured once the dispute was over. Furthermore, their joint activity appeared to be worthwhile: in most places they succeeded in thwarting the masters' plans to impose more wage cuts.[18] Having fought a successful defensive action, their thoughts naturally turned to the prospect of going over to the offensive, using their newly-acquired unity to win better pay and conditions.

Thus, by November 1842, a transformation had taken place which had created the conditions necessary for the formation of the Miners' Association.

Initially, the Miners' Association drew its support entirely from the county of its birth—Yorkshire.[19] After consolidating the union's position there, the general secretary, David Swallow, ventured into the North-East. His campaign in Northumberland and Durham, which began in January 1843, met with immediate success. Within a few months, he had recruited the majority of the miners in what was then Britain's largest coalfield.

In June 1843, with the union's base secure in Yorkshire and the North-East, Swallow turned his attention to Lancashire. In a letter to the *Northern Star*, he stated that he intended to reach Oldham by 27 June; any colliers wishing to contact him could do so through the *Star*'s agents at Oldham or Wigan. Swallow, who was accompanied on his mission by Daniel Thompson, added, 'Brethren, we are entirely strangers to your district'.

The first public meeting seems to have been held at Kersal Moor on 29 July. This was attended by only 150 miners, whose numbers were swollen by the presence of colliery owners, magistrates and constables. The chair was taken by John Lomax, of Radcliffe Bridge. He said the object of the meeting was to form a national union to secure a fair day's pay for a fair day's work. Other speakers took up the same theme. George Harrison said:

> 'He had a wife and five children; and got between eighteen shillings and a pound per fortnight. If he went to the Poor Law Guardians for relief, he would get fifteen pence per head, which amounted to eight shillings and ninepence per week, so that by working hard in a coal mine, he obtained an extra threepence.'[20]

David Swallow mentioned that women working at Earl Balcarres' pits had to drag three tons of coal to the surface for every penny they received.

The poor attendance at the meeting was because some men confused Kersal Moor with Kearsley Moor, near Bolton.[21] So, to clarify the position, a second meeting was held two days later. This was followed by the first Lancashire delegate meeting, held in a large room in Brown Street, Manchester, with representatives from twenty-three collieries present. Its most important task was to map out the membership drive for the forthcoming weeks.

This campaign appears to have been successful. In a report to the national headquarters at Newcastle-on-Tyne, Daniel Thompson said:

> 'Swallow and I are doing wonders in Lancashire ... but we are watched at every step by the creatures of the masters, acting as spies upon our actions. They even get into our sleeping-rooms to listen to our private conversation, but as our work is done openly and above board, we have nothing to fear and defy their malice.'[22]

By this time, Swallow and Thompson comprised part of a team, touring the county to rally colliers to the union. They received assistance from people with contacts and experience: William Grocott and William Dixon, the Lancashire miners' secretary and treasurer respectively, were both prominent Chartists; Henry Dennett and John Berry were both veterans of previous miners' struggles, having belonged to former unions.[23]

Swallow was encouraged by the progress they were making. 'It is now three weeks since our first meeting in Lancashire,' he wrote to headquarters. 'I can say that the work here is going at a far faster rate, than it did in the North.' Repeatedly in his reports to Newcastle, Swallow returned to the subject of the Lancashire miners' poor pay:

> 'I can name several places here (Lancashire) where they do not yet get ten shillings a fortnight—for eleven days' work too! I leave you to judge of the matter when they give no more to a man who

has to break his day's work to attend a delegate meeting and to walk ten miles there and back, than two shillings for wages and travelling expenses of the meeting. I heard delegates declare that they were as well, or better paid, than those who were working at the pit. The colliers in this part of the country live cheerfully on a little oatmeal mixed with boiling water. This is the main food when eaten with buttermilk.'[24]

Lord Francis Egerton employed 1,300 underground workers at his Worsley collieries paying them only a few pence more than if they were unemployed and went to the Poor Law Guardians. When Richard Heathcoat was killed at one of His Lordship's pits, Swallow exclaimed:

'Who would be a collier to be exposed to hourly death; to have his head split in two; his brain dashed out; and all for fourpence a day?'[25]

David Swallow said the poorest paid colliers in Lancashire worked for William Hulton, of Westhoughton. Their plight can be judged from this appeal for more money:

'We ask it on behalf of our once blooming wives, but now hunger-stricken and emaciated wives and children; we ask it on behalf of ourselves who, in many instances, have to descend into the pit without a breakfast, for no purpose but that ye be enabled to grow rich while we sink by slow degrees into a premature grave, victims to over-exertion, nauseas gases and starvation.'[26]

When Swallow went to Westhoughton, he encountered opposition from Hulton and the other coalowners. They prevailed upon the local publicans to deny Swallow a meeting place. At the same time, they provided their miners with free beer and played bowls with them. This display of cordiality, obviously intended to keep their colliers from the clutches of the union organizer, did not daunt Swallow, who proceeded to hold an open-air meeting. Whereupon Hulton came over to stop him. Swallow protested that he was causing no obstruction. Hulton replied that, if he were not gone in a few minutes, he would be sent to prison.

'Have you not threatened the publicans if they allowed us to meet?' Swallow asked. 'And does not your whole conduct prove that you dread the diffusion of intelligence among your vassals?'

'Well, you have no occasion to come here and teach us,' answered Hulton. 'We know more than you do.'

'I do not know quite as much about Peterloo,' conceded Swallow.

'I had to go and work in a coal-pit when very young to help aggrandise such men as you.'

At this, two men who were employees of Hulton, lumbered over to Swallow and said, 'Dun you noa the consequence oa taking' abeawt Peterloo to th' Squoire?' They had a short way with union organizers —they threw Swallow from their midst.

The reference to Peterloo, it would seem, was made by Swallow because William Hulton was the magistrate who, on 16 August, 1819, ordered the Manchester and Salford Yeomanry to charge on horseback into the crowd at a large meeting on St Peter's Field, with the result that there were numerous casualties. The affair went down in history as 'the Peterloo massacre'.[28] If, as seems probable, it was the same William Hulton who was involved in the Westhoughton incident of 1843 as in the earlier trouble of 1819, it would seem that in the intervening twenty-three years he had become no more reconciled to the concept of permitting the right to free assembly.[29]

Notwithstanding the hostility of Hulton and other coalowners, the Miners' Association continued to make members. Even collieries like those at Worsley and Westhoughton—'strongholds of corruption' Swallow termed them—eventually succumbed. A delegate meeting at Ashton-under-Lyne, on 23 October, 1843, heard that 98 lodges had been formed in Lancashire and Cheshire. A further sign of the union's strength in the North-West was that 45 per cent of the eight thousand copies of the *Miners' Advocate* were sold in the region.[30]

Clearly the Lancashire miners, with such a large membership, were in a position to play a leading role in determining the union's national policy. At the beginning of 1844, a debate raged within the ranks on the question of industrial strategy. Conferences at Manchester in January and Glasgow in March heard the contending speakers. Martin Jude and the South Staffordshire delegates argued for a policy of endeavouring to achieve a reconciliation with the coalowners; they repudiated the strike weapon. W. P. Roberts and the miners from the North-East, on the other hand, believed that the only way to gain a major improvement in pay and conditions was to have a general strike. Both sides condemned partial strikes—that is to say, strikes at one or a small number of pits. The moderates, of course, based their opposition on the grounds that they were, *per se*, opposed to men withdrawing their labour. The militants were against partial strikes because they did not want resources frittered away in small, inconclusive conflicts, since, in their opinion, only a general strike could achieve a lasting amelioration.[31]

The false assumption seems to have been accepted by both sides that the union would be able to avoid partial strikes if it wished. It was apparently thought that the Miners' Association would be permitted to grow in size and influence without any outside interference. Nobody appears to have envisaged that the employers, by taking pre-emptive or retaliatory action, might embroil the organization in partial strikes. In fact, many masters regarded the mere existence of the union as an outrage that could not be tolerated. While many coalowners became irate, the existence of a union made the colliers acquire greater confidence. They were unwilling, whatever, the leadership said, to exercise restraint. Rather they wanted to reap immediate rewards from their newly-created unity. Incompatible attitudes were developing among the coalowners and their employees. Simply for the miners to have a trade union disturbed the balance of industrial relations and engendered conflict.

In Lancashire, where wages were exceptionally low, colliers quickly grew impatient. At the beginning of 1844, thousands of men stopped work. They sent representatives to the national conference, then being held at Manchester, calling upon the Miners' Association to support them. This the conference refused to do. Instead it sent two spokesmen, Beesley and Dixon, to St Helens to explain why the union, at that moment, favoured caution.

But caution was not the mood of the Lancashire men. Soon after dispatching its emissaries to St Helens, conference received another demand for immediate action, couched in the following quaint language:

> 'Dear Brethren,
>
> This is to let you know that we, the miners of the Earl of Balcarres, have agreed at a great meeting that the present time is the most suitable for a strike, and are determined to have one, believing the corn is ripe at the present in our quarter; and if you be on our side, we shall be on your side when yours is ripe; and if not, remember that we are determined to withdraw, for it will totally break us up.'

'After reading the letter, the chairman immediately tore it to pieces as a mark of disapprobation with its content.'[32]

The antagonism was mutual. Lancashire strikers thought they were being betrayed; the union should back them by giving financial support. Some men expressed their annoyance by leaving the Association.[33]

But, whatever the national policy, it did not materially affect the course of the dispute. The St Helens miners rejected the offer of a ninepence a day increase. Later, they changed their minds when they heard of the coalowners' plan to introduce Welsh labour. But, equally, the coalowners changed their position: believing that the threat of Welsh immigrants strengthened their hands, they linked the wage offer to the introduction of a yearly bond and renunciation of the union. The following week the first contingent arrived from Wales, and they were met by a thousand strikers, 'who assailed them with stones, brick-bats, and other missiles, severely hurting several of the police'.[34] Strikers captured the Welshmen from the police and, after giving them refreshments, persuaded them to leave St Helens.

The St Helens police force, which numbered eleven, but had been doubled because of the disturbances, was incapable of coping with the situation. It succeeded in capturing several of the strikers who had been prominent in the affray, but had to release the prisoners when an angry crowd surrounded the police station. Only the arrival of seventy soldiers of the 67th Regiment restored law and order.[35]

The St Helens coalowners decided not to make any more attempts to import Welshmen. They were faced with idle collieries and rising coal prices. The only way they could restart production was to make a big concession to the men. Wages were increased by a shilling a day, an advance that represented between a third and a half on prevailing rates of pay.[36]

This big victory for the St Helens miners was repeated, with slight variations, in other parts of the Lancashire coalfield. The rapid rise in wages engendered a feeling of euphoria among the men, a belief in their own power and a confidence in their own organization. Any coalowner who endeavoured to resist their demands was liable to have his colliery shut indefinitely. An indication of their might is given by their ability, in the autumn of 1844, to sustain and win a seventeen-weeks' strike. The *Manchester Guardian* explained, 'The only grievance ever alleged against the proprietors of these works is the employment of one man who refused to subscribe to the Miners' Association'.[37]

The strength of the Lancashire Union contrasted vividly with the position of the union in most other coalfields. In the spring and summer of 1844 most of the district organizations of the Miners' Association had been involved, either as a result of their own impetuosity or the masters' provocations, in damaging disputes. Many of the district unions had been smashed. This was true of the Association's stronghold in the North-East where, after twenty-two weeks' strike, the men were compelled to return to work on the coalowners' terms. Wholesale evictions from their cottages and the mass importation of Irishmen, which created a labour surplus, demoralized the men and wrecked the union there.

The debacle in Northumberland and Durham meant that the centre of gravity within the union shifted from the North-East to Lancashire. Members looked to Lancashire rather than Newcastle for a lead. It seemed logical that the headquarters should be moved to Lancashire. Initially, this was done so that the general secretary could be in closer contact with the editor of the union's journal, the *Miner's Advocate*, which was being published in the Isle of Man to avoid the stamp duty on newspapers imposed in this country.[38] But, eventually, it became clear that a new leadership, with new policies was required. William Grocott and William Meadowcroft, Lancashire county secretary and treasurer respectively, replaced John Hall and Martin Jude, who had been secretary and treasurer of the national organization. All the leadership of the Miners' Association came from the North-East.[39]

Lancashire provided the union with different policies and a different style of operating. Whereas the Newcastle leaders had attempted to get all the coalfields to implement the same policies, under its central guidance, the Lancashire leadership believed in greater flexibility and more local autonomy. The prosperity of the Lancashire colliers had been won through partial strikes, struggles fought in direct contravention of national conference decisions. Therefore, its leaders had no time for those who still wished to pursue the chimera of a national industrial strategy; given the diverse economic and social conditions this was unattainable. Nor did they agree with the views of Martin Jude and others, who pointed to the ruined organization in the North-East coalfield and concluded that all strikes must inevitably be disastrous. By pursuing a policy of partial strikes, concentrating its efforts on one area, picking off the coalowners one at a time, the Lancashire miners had continued to gain more and more wages.

The Lancashire union envisaged the Miners' Association, as a national body, performing five functions. First, it gave representatives from the various coalfields an opportunity to meet periodically and discuss problems of mutual interest. Second, it provided the means whereby financial aid could be channelled to any district that was in severe trouble. (Lancashire contributed generously to the North-East colliers during their big dispute in 1844.)[40] Third, it served as an information service about industrial disputes, the means by which one district contacts another district which then did everything possible to prevent blacklegs from leaving for the strike-bound coalfields. Fourth, it acted as the co-ordinating body for campaigns to secure improved mining legislation—holding meetings, organizing petitions, providing spokesmen for delegation to Parliament. And, fifth, it gave the union's membership legal protection by employing a full-time lawyer.

From 1844 to 1846, the Lancashire union experienced a period of unequalled prosperity. With the industrial expansion of the North-West, the demand for coal rose rapidly, and consequently the demand for labour also rose. The Lancashire union was favourably placed to exploit this opportunity. Miners' organizations elsewhere, although generally debilitated by the unsuccessful strikes of 1844, still possessed sufficient effectiveness to prevent large numbers of men going to Lancashire to work whenever an industrial dispute occurred. So the Lancashire union was able, without any outside interference, to pursue its policy of fighting limited engagements, forcing particular coalowners to concede increases knowing that others, under threat of an indefinite stoppage, would follow suit. In this way, they secured five increases in two years, a rise of half-a-crown a day which, for some men, doubled their income.[41]

The strength of the men's organization, which was in marked contrast to the absence of any equivalent organization on the part of the coalowners, made masters apprehensive and ready to adopt a policy of appeasement. When the union held its large rallies, where the men were usually harangued by W. P. Roberts, it was not uncommon for the employers to send along peace-offerings.[42] Meat, bread and ale thus provided went to make demonstrations more convivial. Some coalowners even gave food and beer so that men at their own pits could celebrate the formation of their local union branch.[43]

This display of generosity should not be misinterpreted. It was an attempt to buy industrial peace. The alternative—an indefinite cessation of work—could be exceedingly costly for the coalowner. It would place his main competitors in the neighbourhood at an advantage, able to benefit at his expense by capturing his share of the market and making profits while he made a loss. Therefore, he had every incentive to reach an agreement with the union.

Cordiality and conflict co-existed throughout the coalfield. Whenever the Lancashire union found an obdurate owner, not prepared to grant the desired wage increase, then the full power of the organization would be deployed until victory was attained. Any of its members found guilty of intimidating a blackleg, or committing acts of violence against him, would have their fines paid by the union.[44] Union funds also supplied strike pay for whatever was the duration of an officially-recognized dispute. Messrs Swire and Lees, the owners of collieries at

Duckinfield and Hyde, paid the penalty for failing to concede four shillings a day: from February to December 1845 their production was stopped.

While prosperity continued for the Lancashire miners until the slump of 1847-8, signs of impending catastrophe appeared much earlier. The first serious setback came with the Bolton strike of 1846. It began in January with the men quite optimistic, expecting the dispute to follow the normal pattern. But a temporary dip in coal prices made some owners, such as the Earl of Bradford, prepared to resist. These employers made plans to acquire alternative labour. As a result, strikers saw newcomers taking their jobs and became increasingly worried. Some of them signified a willingness to return to work on the old terms. Other colliers stayed on strike and finally discovered employment was no longer available. One thousand five hundred men were left without work. This reserve army of labour, coupled with the union's failure to obtain the wage increase, led to the disintegration of its organization in the Bolton area.

The fact was that the Lancashire union was finding it more and more difficult to preserve its own strength when miners' unions in other parts of the country were collapsing. Better conditions in Lancashire than in the surrounding coalfields had a magnetic attraction: the pull of good money proved to be irresistible to some outsiders. While there was a small but steady trickle of men into Lancashire during normal times, which had the effect of depressing the wage rates slightly, when industrial disputes occurred miners from elsewhere showed less and less reticence about working in the county. Welsh colliers, many of whom received only half the pay of their Lancashire brothers, had little sympathy when they heard about strikes in Lancashire which, if successful, would result in the differential growing larger.[45]

The influx of strikebreakers became increasingly serious. As unions in other parts of Britain ceased to operate, there was no apparatus in the various coalfields to counter the coalowners' propaganda, following their agents around as they sought to recruit scab labour. So Lancashire colliers had to rely upon making their own appeals. But appeals to men to show solidarity and brotherhood had diminished effectiveness. It was tantamount to asking them to behave like trade unionists when they were no longer trade unionists. Thwarted and frustrated, the Lancashire colliers fell back upon their final ploy, using abusive language and threats of violence.[46]

The decline of the Miners' Association in the North-West was accelerated by rivalry and disunity within the ranks. Chartism, which had been a powerful force promoting the development of the union from its early days, unintentionally served to hasten the demise.

A Chartist-sponsored body—the National Association of United Trades—caused the friction. At first glance, this may appear strange because initially its relations with the Miners' Association were so cordial. Both bodies had prominent Chartists among their leaders. The inaugural conference of the National Association, held at London in March 1845, was chaired by John Berry of the Lancashire miners. W. P. Roberts sat on the National Association's nine-man executive. From the variegated nature of the county's economy, the Lancashire miners had gained considerable experience of working in harmonious collaboration with workers from other industries. They supported the National Association's aim of making this collaboration more firm and more widespread. Further, the National Association's federal structure corresponded to that of the Miners' Association under the Lancashire leadership.

It was precisely because they were identical in so many respects that the friction arose. Instead of being complementary, the two organizations were competitive. They behaved like shopkeepers, out to enlarge their business at the other's expense. As Britain moved towards a trade depression and other potential 'customers' became less, the rivalry between the two organizations grew more vicious, with character-assassination becoming a frequently-used weapon.[47]

The National Association strived to inveigle W. P. Roberts to loosen his links with the Miners' Association. The *Northern Star* supported this move, arguing that it was a shame Roberts devoted so much of his time to mining matters. This constituted a very serious threat to the Miners' Association. Roberts was more than a mere legal adviser: he was the unofficial leader of the union, the main speaker at its demonstrations, and the man who inspired the greatest fear among the coalowners. When the National Association tried to get Roberts to disaffect, it was endeavouring to rob the Miners' Association of one of its most precious possessions.[48]

In spite of this dispute, the Miners' Association remained remarkably faithful to Chartism. Its general secretary, William Grocott, declared, 'I would rather have my right hand severed from my body before I would be a conspirator against the *Northern Star*.'[49] In the last extant issue of the *Miners' Advocate*, the editor stated:

> 'We believe Mr Feargus O'Connor to be a pure and incorruptible patriot, and a sincere, uncompromising, and devoted friend of the

oppressed working man. The National Land Company, of which he is father and founder, will form a new era in the history of Labour's Redemption.'[50]

Whenever union lodges held dinners, it was customary to have a toast to Feargus O'Connor and W. P. Roberts. In the 1847 general election, the Miners' Association did everything it could to support W. P. Roberts' fight at Blackburn, publishing his election address in the *Miners' Advocate* and encouraging workers to show their support on polling day.[51] Similarly with William Dixon at Wigan: the union combined with local Chartists to give him all possible assistance in the contest.[52]

Indeed, it was the strength of its Chartist connections that ultimately led to the break-up of the Miners' Association. Reeling under the effects of the trade depression of 1848, its final collapse happened because some of its prominent members were arrested when the government started a policy of wholesale repression. In December 1848, William Grocott, along with seventy-five other Chartists, was jailed at the South Lancashire Winter Assizes.[53] On 23 January, 1849, nine prisoners at Kirkdale Jail submitted a memorial to the Home Office, asking for clemency, but nothing came of their plea.[54] These nine included William Grocott (the general secretary), William Dixon (whose premises were used as union headquarters), and James Leach (who was responsible for distributing the *Miners' Advocate* in Lancashire).

The union—or what was left of it—had been beheaded.

Table 3 *Parliamentary returns for the export of coal*

1828	357,864 tons	1835	736,060 tons
1829	371,271 tons	1836	916,868 tons
1830	504,419 tons	1837	1,113,610 tons
1831	510,831 tons	1838	1,313,709 tons
1832	588,446 tons	1839	1,449,417 tons
1833	634,443 tons	1840	1,606,313 tons
1834	615,255 tons	1841	1,848,294 tons

Table 4
Income from Coal Duty on coal entering the Port of London

1840	£86,996
1841	£93,851
1842	£86,464
1843	£89,951
1844	£85,747
1845	£108,915

Source: The History of the Coal Trade, Picton Library, Liverpool.

Table 5 *Export of Coal from Liverpool*

1833	£50,561
1834	£59,078
1835	£61,542
1836	£90,042
1837	£95,188
1838	£95,648
1839	£103,630

Source: The History of the Coal Trade, Picton Library, Liverpool.

CHAPTER THREE

Changes in the Mid-Century

It is a perplexing task, fraught with difficulties, to assign a definite date to the collapse of a colliers' union. The symptoms of decline—a loosening of ties, a lessening of mutual activity—happen gradually, and it becomes almost an arbitrary judgment when to affix the death certificate. In an important sense, the union never dies. By the middle of the nineteenth century, organization among miners had reached a sufficient level of maturity to ensure that, whatever happened nationally at individual collieries men retained some semblence of solidarity. This can be seen in Lancashire. The Miners' Association is supposed to have perished in 1849. The end of this national body, did not mean trade unionism as a whole collapsed. In 1850, the journal the *Red Republican* reported a strike of miners at Royton. The news, which appeared in the same issue that published 'The Communist Manifesto' for the first time in English, said the men had been on strike for some time and had been well supported financially.[1]

In the opinion of some colliers, such strength at local level remained inadequate. An attempt was made to re-create a national union. In 1851, a conference formulated a fresh constitution, published it in the magazine *Friend of the People*, and appealed to all colliers to send subscriptions to Newcastle upon Tyne.[2] Its efforts proved to be abortive, perhaps because it strived to resurrect the discredited North-East leadership of Martin Jude. Lancashire colliers just continued to maintain links with other coalfields simply by correspondence, and they used David Swallow, that well-known veteran of the Miners' Association, as their secretary.[3]

In other ways, the imprint of the defunct national organization was firmly fixed on the county union. Its constitution was modelled on that of the Miners' Association, with large chunks taken from the original draft passed by the 1842 Wakefield conference.[4] Then, also, the title was derivative: the 1850s union called itself the Lancashire and Cheshire district of the Miners' Association, an indication that colliers in the two counties considered the national union dormant, not dead.

But there were important differences from its predecessor. The earlier union had not bothered providing friendly benefits. Its main emphasis had been on industrial struggle. By contrast, the latter union not only provided this service but warned its members about becoming involved with rival friendly societies, which sometimes defaulted in their payments. Another difference was that it had written into its rules that county meetings would only be held irregularly. This may seem somewhat strange. Its enactment was to ensure that no county leadership arose, thereby taking power from local leaders. Towards the same end, officials of the county organization received small payment —the finance secretary ten shillings a fortnight, the treasurer five shillings a fortnight—thus making certain that any full-time officials would play a minor role, appointed by, and answerable to, the local lodges.

According to the rules, the functions of the county meetings were to elect officers; to determine itineraries for its agents; to inquire into market conditions; and to investigate problems of health and safety. The county meetings could also impose fines, ranging from two shillings for slander to ten shillings for unauthorized striking. This latter point denotes a fresh departure. Under the 1852 rules, when a dispute arose at a colliery, five-sixths of the membership must vote in favour of strike action before it could take place. Even if there were this majority, their decision had to be reported to the district meeting, which might propose arbitration as an alternative course.

Safeguards against strike did not stop them happening. There was a particularly vicious struggle at Wigan in 1853. The colliers demanded a twopence in the shilling increase; the owners refused; soon five thousand men were idle. A week later the same thing happened to the textile workers. As a result, the number of strikers in the town was swollen to thirteen thousand.

After seven weeks, stomachs were empty and tempers frayed. A local newspaper reported, 'Hundreds of men, women and children—the latter especially—are making a sort of house to house visitation in the borough, and are also extending their applications for miles around the neighbourhood ... Inhabitants of houses in the outskirts of the borough and neighbouring villages have been somewhat startled by numbers of colliers asking for alms in a peremptory manner and, where money or bread was refused, threats have been made which excited much alarm.'[5] Magistrates resolved to try and stop the begging, a move that only served to irritate the miners. Exasperation turned to anger when Lord Crawford gave his miners notice to quit their cottages and brought in strikebreakers from Wales.

The spark that finally caused the explosion came on Friday, 28

October. Coalowners were meeting in the Royal Hotel, Wigan, discussing the men's claim, while thousands of colliers stood outside, awaiting the result of their deliberations. When news reached the crowd that the owners rejected their demands, the strikers surged forward in anger. Maskell Pease, manager of Earl Crawford's collieries, fled up Market Street, taking temporary refuge in a chemist's shop and later in the Dog and Partridge Hotel. The major, chief constable, nine police officers (the entire Wigan force) and about twenty specials arrived as the crowd was making an onslaught on the hotel. The civic authorities were greeted with potatoes, apples, oysters and tomatoes. A magistrate was hit by a stone and stunned while an alderman, standing next to the mayor, was struck with a piece of wood. 'The volley of missiles was thick and heavy, and all the special constables and some of the regular officers disappeared ... At last only the mayor, chief constable and two police officers were left to face the crowd'.[6]

For five hours, the mob controlled the town. Public houses, the town hall, post office, grammar school were all seriously damaged. People even attacked the police barracks in Millgate, where most of the constabulary were hiding. Men with bludgeons—wood ripped from market stalls—roamed around destroying or stealing whatever they chose. Only when Captain F. S. Gerrard and the Lancashire Hussars arrived was law and order eventually restored. Even then an attempt was made to destroy the engine-house of Moss Hall colliery with a home-made bomb, which blew off the roof and shook nearby buildings.

The Wigan riot aroused an outcry in the national press against 'the shameful negligence of the Authorities', the failure of the mayor to telegraph for help until almost three hours after the disturbance began, and the mistaken belief that a turbulent town of 35,000 could be controlled by nine policemen (three first-class constables at 19s. 3d. a week, three second-class constables at 18s. 1d. and three third-class constables at 17s. 6d.).

Main criticism, however, was directed at the miners, as in the *Manchester Guardian*, editorial:

> 'It would sometimes appear to be thought—and probably the the loose talk of demagogism has designedly strengthened the error —that the lowest labour in the intellectual scale is that which society can least afford to part with. It is probable, for instance, that the Wigan coal-miner is possessed of some idea of the permanent value of his order. Without those large bones and thick sinews of his, he thinks that every art and business must come to a standstill: and

from thence it is but an easy transition for him to suppose that his class will always retain the power of reducing the world to capitulation.'[7]

The *Manchester Guardian* regarded the Wigan collier as 'a monstrous nuisance', 'a pitiable mistaken creature'.

The Wigan strikes occurred against the background of general industrial unrest throughout Lancashire. As W. W. Rostow points out, trade expansion, accompanied by a large volume of long-term investment, created virtually full employment.[8] This transformed the situation in the labour market. Workers realized that conditions were propitious to push for wage advances. Spinners, weavers, factory workers, even the police, struck in order to secure increases. The most hard-fought industrial battle took place at Preston. Normally a low-wage area because of the influx of immigrants and agricultural workers, a call for a ten per cent increase led to a thirty-eight-week lock-out involving more than 20,000 workers.[9]

Ernest Jones, the Chartist leader, understood that, if workers were to have the greatest possible effectiveness in the struggle that were taking place, it was vital for solidarity to be established. In the article where he first advanced the idea of a Labour Parliament, Jones wrote:

> 'Look at the colliers of Wigan and the men of Preston! There is a case in point. Not a link of action in common exists between the two! Yet what a mighty accession of strength each might gain from the other!'[10]

The Labour Parliament, convened at Manchester in March 1854, did not accomplish its aim of uniting colliers, textile workers and the rest. A row started early in the proceedings when David Swallow rose to address the meeting. The chairman objected on grounds that he was not a delegate, whereupon Swallow made it clear that, unless he was allowed to speak, the representatives of the Lancashire and Cheshire miners would leave. They favoured a national association of trade unions, but were not prepared to embroil themselves in an organization which, they feared, would make them 'tools for political and party purposes'. When it was ruled that the miners could remain only as visitors, Swallow and his colleagues noisily left the hall.[11]

This incident shows how much the miners' attitude to Chartism had changed. In the 1840s the Miners' Association had many leading members who were Chartists. Their journal highly praised Feargus O'Connor and W. P. Roberts. Even Swallow himself, when general

William Prowting Roberts, the miners' attorney, who fought many legal battles in Lancashire. This portrait of him was issued by the *Northern Star*.

(a) The frontispiece of the official organ of the Miners' Association. It was published from 1844 to 1849.

(b) Thomas Halliday, president of the Amalgamated Association of Miners. *The Beehive*, a radical journal, published this drawing of him when he sat on the Parliamentary Committee of the TUC.

secretary of the national union, said that the Charter must be won at all costs.[12] The Chartism had been a powerful and useful force, helping colliers to strengthen their union. Later, once it had become divisive, miners discarded Chartism. By the 1850s, Chartism was much weakened and had little to offer. In 1854, the Wigan colliers won their wage increase through their own efforts. This encouraged confidence and self-reliance; there was little need to look elsewhere for assistance. The miners' leaders went to the Labour Parliament, saw the scantiness of union representatives, and probably thought no economic advantage would accrue from establishing direct links. Complacently, they decided it was best to continue in their old, familiar way—at least, it was gaining results.

His resistance to change, the local separatism of the Lancashire miner, had a powerful economic basis. Usually he worked in a tiny pit, with a few mates. Other friends would work at collieries dotted about for many miles around. So the dispersion of miners and the small-scale of production made it very difficult to create a centralized organization. With his mental horizon limited to the problems of his own particular pit, it took great effort for him to think in terms of county affairs. A nearby town or village, where he frequently talked and drank with men from other pits, served as a centre for his trade union activities as well as his social life. Hence union organization in Lancashire was distinctive: it consisted of a conglomeration of small, separate unions, independent of each other and imbued with a spirit of local patriotism, which were linked loosely in a county federation.

Small-scale production lingered on longer in Lancashire than in most other coalfields. An 1851 survey of the Wigan district showed that 58 of the 100 collieries had a daily output of less than 100 tons.[13] Likewise Leigh had a similar pattern; it seventeen pits ranged from the smallest, employing six men and producing only fifteen tons daily, to the largest, which employed 52 men and had a daily output of 208 tons.[14] What applied in Leigh and Wigan appears to have been true in the county as a whole.

Yet times were changing: the days of the small men were numbered. For, as the Wigan survey indicates, the large producers were beginning to dominate the market. Of the 100 Wigan collieries, six produced a quarter of the coalfield's output, nine produced a third. When analysed in terms of companies rather than collieries, the tendency towards concentration becomes even more marked. J. Dickinson, the mines inspector, talked in 1854 about the immense difference in size of Lancashire collieries: 'The smaller ones employ only a few hands, and the larger, from one thousand up to fifteen hundred persons. The output of coal per colliery varies from a few tons to a third of a million tons per annum.'[15]

The large firms constituted the dynamic sector of the industry. It was they who had the capital for expensive machinery that deeper workings demanded. While small owners foraged around in shallow mines and inclined planes, usually with a meagre output per man, the large owners were acquiring managements with the skill to handle large numbers of men and large quantities of coal. Technical expertise was vital if disasters, more likely with deep mining, were to be averted. Also, it was required if extensive waste—the loss of up to 40 per cent of the coal—was to be avoided.[16] Even the method of computing mineral royalties necessitated considerable geological and mathematical ability:

> 'In Lancashire and adjoining counties it is customary to let or sell the coal on the principle of so much per foot thick per Cheshire acre (10,240 sq. yds.). The quantity worked is ascertained by making a careful survey of the workings each half year and taking the average thickness of the seams. This method of computing Royalties requires very nice management on the part of the surveyor.'[17]

Nor did the matter end there. Mineral royalties were normally paid irrespective of whether or not the coal was mined. A company which, owing to misfortune or mismanagement, found itself having acquired mineral rights and not used them, would have to pay big sums in 'dead rents'. To avoid such calamities, owners had to make a very careful assessment of their future production capacities and market trends. One of the reasons for the frequent failures of small collieries, as well as the co-operative ventures, was that they lacked officials with the necessary knowledge.[18]

The giants of the industry were able to take a more olympian view. Instead of reacting to events, as their smaller brethren tended to do, they wanted to influence the course of development for their own benefit. Consequently, they were the driving force behind the formation of the South Lancashire and Cheshire Coal Association in 1855. Their immediate concern seems not to have been resisting attempts of the workers to win higher wages, but rather to influence parliamentary legislation. Clearly, with the Mines Act 1850 regulation of collieries was here to stay, and coalowners themselves had a vested interest in it. While they were concerned that any law did not interfere too much with production, they still wanted to avoid pit disasters, which frequently resulted in costly damage to the mine, a demoralization of the labour force, and a loss of life to both management and men. The death toll was often high among officials responsible for safeguarding lives and property. In his article, D. Anderson shows the remarkable fact that all the general managers at Blundell's collieries between 1854 and 1896 were killed down the pit.[19]

Although major disasters provoked the most comment from press and public, it was the day-to-day mining operations that caused the greatest carnage. Accidents involving a few individuals or even a single man were, when considered along with working conditions themselves, responsible for exacting the biggest toll. That coal mining was not conducive to health and longevity was conclusively proved by Dr Herbert Mackworth's article, 'On the Diseases of Miners',[20] which contained the following table:

	Coalminers	Agricultural Labourers	Other Labourers
av. age of living	26·1	34·2	34·0
av. age of commencing	11·2	11·3	14·6
av. age of leaving off work	39·7	53·6	50·6
No. of years of labour	28·5	42·3	36·0

Dr Mackworth's survey gave the national picture; in some respects, the position in Lancashire was worse. Dickinson reported that 'the steepness of the seams in many of the mines ... and the extent to which underground machinery is applied render the working of the mines more than usually dangerous'.[21] Gassy seams also aggravated the problem. Dr P. E. H. Hair estimates that, by 1851-3, Lancashire and Cheshire had a death rate from explosions that was twice the national average; overall, taking all deaths from mining operations, Lancashire was a third above the national average, and, except for the Midlands, was the most dangerous district.[22] One life was lost per 46,010 tons produced in Lancashire as against one life per 54,822 tons for the United Kingdom.

Both management and men could hardly fail to take cognizance of this appalling accident rate. Significantly, while miners from other coalfields did not bother to send representatives to testify before the Select Committee on Accidents (1854), Lancashire dispatched Swallow and Tetlow. Likewise coalowners sought to tighten safety regulations.

The Mines Act (1850) permitted them to introduce by-laws, rules to fit special local working conditions, and Lancashire masters used this right to the fullest extent. In Wigan, coalowners even tried to use this power, with the severe penalties that could be imposed for infringements, to make colliers 'attend Divine Service at least once every Lord's Day' and to punish those coming to work 'on Monday morning dirty or with an old beard'. When the Home Office heard of these provisions, the Wigan coalowners were told they would have to be expunged since they did not come within the meaning of the Act.[23]

Even where stringent and carefully devised regulations for underground working were introduced, they availed nothing if, through lack of knowledge and indifference, they were just disregarded. Dickinson, the mines inspector, told the Select Committee on Coal Mines (1852) that Ince Hall colliery had one of the best sets of regulations and in his 1852 annual report he included its regulations, hoping other collieries would model their own rules on them.[24] But his proselytizing received an embarrassing rebuff when, in March 1853, an explosion at Ince Hall killed fifty-nine men. The immediate cause of the disaster was that the furnace, necessary for proper ventilation, had been permitted to go out. But the subsequent inquiry revealed that the regulations had been broken in numerous ways. Dickinson commented:

> 'That an instance of better rules and worse discipline he had never met with, and, for any practical use, there might as well be no rules. They were all broken by the very officers who were supposed to enforce them. They all seem to have permitted smoking, and also blasting with gunpowder ... No qualified person was appointed to fire the shots; and all precautions were systematically broken.'[25]

In most accidents an important contributory factor was lack of education. The high Lancashire casualty rate should not be attributed, as Dickinson did, to the introduction of machinery—in the long-run it resulted in a reduction in the accident rate—but to the putting of potentially dangerous equipment in untutored hands and under inadequate supervision.[26] The Select Committee Report of 1850 alluded to the defective state of mine management in Yorkshire, Lancashire and North Wales. Mentioning this, R. N. Boyd, himself a mines inspector, stated that, 'Many of the accidents on record may be traced directly to ignorance on the part of persons in authority ... the practice of entrusting the management of dangerous collieries to incompetent people is highly blameworthy.'[27] A similar opinion was expressed by David Swallow to the Select Committee on Accidents (1854). While he considered faulty management responsible for most accidents, he conceded that sometimes the men were to blame. The root-cause of accidents was ignorance, and to overcome it, said Swallow, the miners were prepared to pay twopence a week for education so long as they were given representation on the school boards.

Lancashire coalowners saw the problem in two parts. First, there was need to make provision for the proper training of management. Second, there was the need to make better provision for the ordinary miner's children. As a step towards these goals, the Wigan Mining and Mechanical School was formed in 1857. According to one writer, the school 'covered a larger range of mining instruction, and dealt with more mining students, with greater advantage, than any other mining school in the United Kingdom'.[28] Besides the obvious advantages that would be derived from greater technical knowledge, the coalowners hoped that general education would have other beneficial effects. It might inculcate a more responsible attitude among the miners, a greater respect for authority, a lesser inclination to destroy private property and indulge in industrial strife. Therefore, some coalowners felt they had a personal interest in education and supplemented existing provision. In 1854, their assistance was as follows:

Name of colliery

Gidlow & Swinley. Owner provided the building and some books. Also £20 p.a. towards teacher's salary. Each child paid twopence weekly to the teacher as well.

Bickershaw. Maintained by money from the owner and children.

Haydock. Maintained by money from the owner and children.

Astley & Tyldesley. Owner made small contributions; could send eight children to Astley Free School.

Burnley, Habergham, Marsden and Padiham. School maintained by John Hargreaves, coalowner, plus voluntary support.

Hopwood. Several schools 'greatly supported'; 'many men and boys attend'.

Halstead. School built by coalowner. Maintained by him and voluntary contributions. 84 boys and 78 girls attended.

Ravenhead. Maintained by money from the owner and children.

Worsley. Wholly supported by the Bridgewater Trustees. 800 pupils.

Clifton. Maintained by Trustees of Ellis Fletcher.

In comparison to North Wales, where coalowners provided no

schools, the position in Lancashire was good. Yet, even in Lancashire there remained much opportunity for improvement. Dickinson remarked, 'Colliers, as a body, are perhaps less alive than most classes of skilled workmen to the importance of education. Parochial and national schools are numerous, and may be reached almost invariably within two miles of the colliers' dwellings, but few miners' children are said to attend.'[29]

The basic attitude of miners did not change rapidly. Indeed, from a newspaper report in 1868, it appears that there may have been a deterioration: 'Mr Knowles, ex-Mayor of Wigan and an extensive colliery proprietor, states that out of a thousand men there were only five or six who could write their own name, and they were chiefly aged persons.'[30] Corroborative evidence comes from the ninth report of the Schools Inquiry Commission (1868), which describes the apathy about education prevalent in such places as Blackrod:

> 'Now Blackrod is one of the richest coal districts in Lancashire. It is a miserable place. Its so-called grammar school, founded in the time of Queen Elizabeth, is full of the children of miners who come from the national school for the sake of the clothing charity, and leave school altogether for the pit at eleven or twelve years of age.'[31]

Because of the high demand for child-labour, parents sent their children into the pits as soon as possible. They respected education, but valued money more. Economic necessity was paramount, a point of view that was shared by the Commission appointed to inquire into the State of Popular Education in England:

> 'The peremptory demands of the labour market, if the wages of the child's labour are necessary, either to keep the parents from the poor rates, to relieve the pressure of severe and bitter poverty, it is far better that it should go to work at the earliest age at which it can bear the physical exertion than that it should remain at school.'[32]

The Commission realized the economic pressure that was exerted on parents; it appears to have overlooked the economic pressure exerted on employers. For when it talks of the need to improve educational standards, it would have nothing to do with compulsion; rather it looks to 'right-minded employers' to use their 'moral influence'. In the coal industry, where employment of child labour was very profitable, the moral persuasion would be minimal. Indeed, it might be in the opposite direction. For instance, the coalowners successfully opposed the inclusion of an education clause in the Mines Act 1860. When the colliers sent twenty-five petitions, purporting to have the signatures of 300,000, which asked, among other things, for better educational facilities, the coalowners dispatched counter-petitions to Parliament.[33]

Alexander MacDonald, who worked for several years in the 1850s as a schoolmaster, claimed that there was no class in the community which had done more than the miners to influence public opinion in favour of improving the system of education.[34] While this seems a rather exaggerated claim, the miners' leaders did undoubtedly do their utmost to encourage educational endeavour. They realized that, if colliers were to be pulled out of the squalor, degradation and misery that formed part of their everyday life, they would have to develop a new attitude, fresh values and greater self-respect. At considerable personal sacrifice, many leaders had taught themselves the rudiments; because of the immense effort this involved, they prized their intellectual attainments and wished others to follow their example.

Union journals always reported favourably on any attempts at self-betterment. One instance was when *The British Miner* quoted an article entitled 'Lancashire Men at School' from Charles Dickens' magazine, *All the Year Round*. Dickens, who took an interest in education in the North-West and was president of the Union of Lancashire and Cheshire Institutes in 1858, described a strange school with 380 pupils of all ages: 'The benches and desks are occupied by rows of grown up men and grey-haired students ... You saw men clasping their bald heads as if to keep them from splitting over sums in addition.' One of the duties of monitors was to stop students from falling asleep; 'sleep not being allowed in school hours, and brain culture having a tendency to produce somnolency in some dispositions'. There were, however, no discipline problems: 'These old boys are good boys.'[35]

Physical tiredness, a condition not conducive to intellectual effort, may have been the underlying reason for the failure of most attempts to involve Lancashire miners in educational enterprises. Charles Hindley, M.P., tried to interest an Ashton-under-Lyne collier in the local Mechanics' Institute. The miner replied 'that he was no mechanic, and could not be persuaded to interfere with the trade of another body of his fellow men'.[36] Thomas Hogg, secretary and lecturer of the Lancashire and Cheshire Union of Mechanics' Institutes, called attention in his 1849 report 'to the special problems of areas where illiteracy was greatest, as in the coalfields'.[37] In her history of the Mechanics'

Institutes, Mabel Tylecote refers to the difficulty of promoting activity in mining localities:

> 'The Lancashire development was concentrated in the manufacturing districts. Particularly noticeable was the lack of provision in the Wigan coalfield, where standards of education were low. The institute in Wigan, a town which had a population of 25,000 in the 'forties, was reported at that time to exist only in name, and only because of the support given by R. A. Thicknesse, M.P.'[38]

The same pattern appears to have prevailed throughout the Lancashire and Cheshire coalfields, with large towns like Accrington having no Institute whatsoever. Generally, although there were exceptions, the standard of education was low. This tended to re-enforce the cultural and social deprivation of these districts.[39] Colliers' desire for education sprung from a realization of their own inadequacies, a longing to transcend their narrow existence. But they were enslaved by a harsh reality of prolonged and arduous toil that made study extremely difficult.

One consequence of this lack of education, injurious to coalowners and miners alike, was that it increased the problem of reaching a reasonably high standard of technical efficiency in mining operations.

A continual source of anxiety to all connected with the Lancashire coal industry was the vulnerability of its markets to outside competition. Whereas most coalfields were assured of selling coal to other industries located in the immediate vicinity, thus being guaranteed a certain sale on their home ground, Lancashire had no such easy sales. The development of a canal and subsequently railway network, probably more abundant in the North-West than in other parts of Britain, meant that Lancashire could receive supplies from many other places. The Manchester coal market, for instance, was accessible by road, rail and canal to coal from Yorkshire, Derbyshire, North and South Staffordshire as well as Lancashire itself. Consequently, Lancashire coal had to remain competitive. Any increase in price, disproportionate to that of its rivals, might rob it of its home market.

The same applied to coal sold outside the county. In 1852, this constituted about a million tons from a total Lancashire output of 8,255,000 tons.[40] The main outlet for exports was Liverpool: 105,952 tons was sent by coastal ships and 277,642 tons went to foreign countries. In most of these markets Lancashire encountered stiff competition from Cumberland and Welsh coalfields. Likewise in the fast-growing London market: Lancashire had to face competition from every other British coalfield. The total amount of coal entering the Metropolis by land carriage rose from 60,849 tons in 1848 to 414,917 tons in 1852, a sign of the growing importance of the railways.[41] Of this last-mentioned figure, about 50,000 tons came from Lancashire by rail.[42]

Lancashire's sensitiveness to national trends made the county's coal-owners show a greater interest than the average coalowner elsewhere in the future of the industry. It may not be surprising, therefore, that when a national employers' organization was formed in 1855—the Mining Association—it should appoint Maskell Pease as its secretary and operate from his offices in King Street, Wigan.

The same propensity was discernible among the men. Realizing their own well-being was inextricably tied to that of miners in other places, Lancashire miners were always to the fore in any scheme to form a national organization. This applied in the 1830s, the 1840s, and again when the revival came in the 1850s. Far too often they had seen their efforts ruined by outside influences.

Yet, this created a paradoxical situation. For, while the Lancashire men clearly saw the need for a national union of mineworkers, at the same time within their own county they continued with their local unions in discord and disarray. It was easier, say, to get the Hindley miners to agree to the principle of common action with miners throughout the rest of Britain than it was to get them to co-operate with the miners from neighbouring Wigan, with whom they were in competition for members.[42]

In 1858, the Lancashire and Cheshire miners, although disunited themselves and without any county organization, nevertheless took the initiative in convening a national conference at Ashton-under-Lyne. This conference was important because it inaugurated a fresh campaign for better legislation. It was attended by delegates from Scotland, Yorkshire, Staffordshire and a few other places. To Alexander MacDonald, the conference was the first national one he attended, and he regarded it as the assembly that did the preliminary work that led to the passing of the Mines Regulation Act 1860:

> 'It was in 1856 that I crossed the Border first to advocate a better Mines Act, true weighing, the education of the young, the restriction of the age to twelve years, the reduction of the working hours to eight in every twenty-four, the training of managers, the payment of wages in the current coin of the realm, and many other

things too numerous to mention here. Shortly after that, bone began to come to bone, and by 1858 we were in full action for better laws.'[43]

The significance of the Ashton-under-Lyne conference, said MacDonald, was that it 'gave the first signs of life' and indicated that the miners were 'determined that things would not remain as they were'. But an equally significant consequence of the conference was the emergence of Alexander MacDonald as a leader of national stature. A former student of Glasgow University, he was a fluent speaker with moderate views. He was able to establish a rapproachement with spokesmen of the government and employers that leaders with extreme views would have been unable to achieve. Hence MacDonald's influence on mines legislation in the 1860s and 1870s was immense.[44]

Yet, while the Ashton-under-Lyne conference laid the foundations for these future achievements, the year 1858 only served to underline the inadequacy of existing county arrangements in Lancashire and Cheshire. There was a general campaign for a wage increase. Some owners quickly conceded it; others refused. Once the agitation was over, the owners who had made the concessions withdrew them, and wages generally slid back to their previous level. It became clear to the men that, unless they had some means of compelling recalcitrant coalowners, usually the large ones, their efforts would always be fruitless.

To talk in terms of a highly centralized county organization was a too revolutionary concept, conflicting too much with traditional practice to stand any chance of succeeding. An alternative—was to endeavour to cajole individual lodges to combine to form district associations. The first attempt to do this was made on 13 February, 1860, when thirteen lodges in the Wigan area formed a district committee.[45] They jointly campaigned for a pay increase but, once achieved, the organizational question appeared to recede; no attempt was made to hold any further meetings. The Lancashire cotton famine, and the ensuing depression in the coal trade (1861-2), revived the issue once more. In splendid isolation, individual lodges had been powerless to resist the downward pull of wages.

In an atmosphere of desperation, colliers convened a public meeting. Two cuts to their wages, both of ten per cent, had been made and a third was pending. After deductions, many miners earned two shillings and three pence a day or less. Subsequently, William Pickard described the situation:

'He recollected that he was working in the night when a notice for a 10 per cent cut was placed upon the office doors. His heart almost failed him, for he knew at the time that he could not get enough to support his wife and family, letting alone the loss of 10 per cent. He had a meeting of the men, at which 2s. 10d. was raised, and the bellman paid to call a meeting, which was held at Scholes, and they could scarcely get a chairman.'[46]

Samuel Rigby, who took the chair, proposed that each pit should appoint two delegates. Then they could discuss the issue among themselves and with the masters. This was agreed. A week later, a second public meeting was held in Wigan Market Square to consider the state of negotiations. Pickard placed a compromise formula—the acceptance of a five per cent cut—before the meeting, but the miners overwhelmingly rejected it. Then Pickard and other spokesmen returned to the master to report that the men remained unyielding. The expiry of the wage-cut notice came on 12 September, and no agreement had been reached by that date. Perhaps the coalowners assumed that the men's disunity, lack of funds, and generally depressed state would compel them to acquiesce; there would be plenty of grumbling but no action. If this was the case, then the employers must have been shocked when practically all the miners of the Wigan area struck rather than accept the new pay rates. By the evening of 12 September, the coalowners had capitulated and withdrawn the proposed reduction.

The news was greeted with jubilation by the colliers. Between two and three thousand men soon joined a newly-formed union—the Miners' General Sick and Provident Benefit Society. In order to capitalize on its sudden accession of popularity, the Society held numerous recruitment rallies. First, it secured solid support in the town of Wigan itself. Afterwards it enlarged the scope of its activities, holding membership drives in the surrounding villages, such as Aspull, Hindley and Pemberton. Well-attended public meetings aroused feelings of solidarity, boosted morale, and gave a sense of confidence to colliers living in isolated communities.

In all their speeches, union leaders counselled the need for moderation and restraint. 'They had no desire for a head-on clash' with the masters. Pickard proudly told colliers that the Mayor of Wigan, John Cross, himself a coalowner, had joined the Society.[47] Such examples of mutual trust and understanding were to be encouraged, said Pickard. He likened the previous union to 'a ship which was wrecked because of the carelessness of the crew'.[48]

One obstacle encountered in recruitment drives was the poverty of miners. They found it extremely difficult to pay the subscriptions. So union spokesmen tried to induce them to economize on beer and tobacco. Wives were told that, if they let their husbands pay a shilling to the union now, they were likely to receive five shillings back from the employers later. But the impact of these arguments were lessened when the Society's secretary, James MacDonald, indiscreetly confessed he was too poor to pay his subscription.[49]

James MacDonald's personal plight worsened. Towards the end of 1862, he only worked four days in five weeks. He had to sell his mining-gear and pawn his belongings. When an opportunity to work did ultimately arise he did not have the equipment, and had to borrow five shillings for a miners' lamp.

To escape from the grip of poverty, MacDonald absconded with the union funds. These were kept in the publican's bedroom at The Horse and Jockey. All monies were in a box, bolted by three locks, which could only be opened with three different keys. These were in the hands of the Society's treasurer, assistant treasurer and president. But MacDonald prized open the locks, took £125, and dropped it into the lap of his wife, waiting in the road outside.

When the absence of MacDonald and the money was discovered, it was deduced that he must have fled to his native city, Glasgow. William Pickard, accompanied by a detective sergeant, journeyed to Scotland and soon made an arrest. A large crowd of miners waited, in the rain, for the prisoners' return to Wigan. They jeered loudly when MacDonald and his wife, under escort, left the railway station. The money was retrieved. James MacDonald was sentenced to ten years' imprisonment and his wife, who had suffered a miscarriage, to eighteen months.[50]

The Wigan union was fortunate to recover its money. Had the incident happened a few years later, after the legal decision in the case Hornby v Close (1866), it would have had no redress. For that was the case in which the Boilermakers' Society sued the treasurer of the Bradford branch in order to recover £25. The Boilermakers' officials believed the Friendly Society Act of 1855 afforded protection for their funds and enabled them to gain a successful verdict against any defaulting officer. But the Lord Chief Justice ruled that trade unions were outside the scope of the measure, being still illegal, although not criminal, organizations owing to their tendency to act 'in restraint of trade'.

In spite of MacDonald's defalcation, confidence in the Wigan district union was not undermined. It continued to grow and prosper. In October 1863, William Pickard, who said he was working for nothing as he possessed property, reported that many local sick clubs were joining.[51] Twelve months later, there were 112 lodges with 4,258 members. By December 1865, after obtaining a number of wage increases, union leaders boasted that wages were within ten per cent of their highest-ever level and calculated that, since its inception, the union had won £350,717 for its members in Wigan.[52] A secure and enduring foundation had been laid.

Table 6 *Statistics relating to Coal 1859-1861*

Total British output in 1860	:	80 m tons
Consumption in 1860	:	10 m tons for gas-making
		15 m tons for iron
		6·7 tons for export

Consumption per thousand inhabitants in 1860

Britain	2,500 tons
Belgium	1,200 tons
Germany	470 tons
France	350 tons

	1859	*1860*	*1861*
Lancashire			
No. of collieries	381	371	373
Tonnage of coal	10,650,000	11,350,000	12,195,750
Cheshire			
No. of collieries	35	35	39
Tonnage of coal	700,500	750,500	801,570

Lancashire exported 590,128 tons out of a total of 6,788,060 tons of British Coal exported in 1860.

Sources: Robert Hunt, Mining Record for 1860.
Robert Hunt, On the Mines, Minerals and Miners of the United Kingdom, *Journal of the Society of Arts*, December 1862.

CHAPTER FOUR

The Era of the District Unions

A strong union in Wigan would be practically useless if there were disorganization elsewhere. So it was clear to William Pickard and others that, once a secure foundation had been laid in Wigan, they must build organizations on the same model in other parts of Lancashire and Cheshire. The first exploratory foray was in St Helens. It culminated in a meeting at the York Hotel, Parr, on 3 July, 1863, which resolved to set up a St Helens Association.[1] A month later, Pickard journeyed to Bolton where, after speeches from Halliday and himself, the Farnworth and Kearsley District was formed.[2]

Renewed life in these areas helped to stimulate activity in Ashton-under-Lyne. There, and in the Cheshire coalfield, an embryonic organization had survived during the lean years. It was led by Allen Tetlow, whose business premises at Hurst Carr served as a centre from which he worked as miners' agent throughout the 1850s. It seems likely that he played a big part in organizing the national conference at Ashton-under-Lyne in 1858.[3] By 1863, with the general upswing in union activity, it appears that Tetlow played a similar role to Pickard. At Oldham, Ashton-under-Lyne and in Cheshire, he was responsible for re-activating links which had remained dormant and district associations that had gone to sleep.

Six district associations existed in Lancashire and Cheshire by 1864, each with its own particular problems.[4] The Worsley Miners' Association, for example, reflected the drive towards local separatism inherent in Lancashire. With only five lodges and a few hundred members, it could hardly hope to be a strong, viable organization. A more logical arrangement would be for it to form part of the much larger Farnworth and Kearsley District. But that was precisely what the Worsley miners did not want. So, on 20 August, 1863, three days after the formation of the Farnworth and Kearsley District, the Worsley Miners' Association was created.

The Ashton-under-Lyne District remained aloof for a different reason. Whereas Worsley was prepared to participate in general Lancashire activities so long as it was permitted to preserve its separate identity, Ashton-under-Lyne seemed content to maintain a strong district organization with the minimal links with the other Lancashire unions. Tetlow, whose associations with the coal mining industry went back forty years, frequently addressed public meetings throughout the two counties. In this informal fashion, personal contact was established; hardly a substitute for the lack of a regular, systematic co-ordination of activity. The Ashton-under-Lyne and Oldham District appear, regardless of elsewhere, to have gone on their own separate way.[5]

In St Helens the problem was the legacy of the past, the struggle of the new against the old form of unionism. H. Davidson, the miners' agent, who had held a similar position in the 1840s, exemplified the latter. He visualized his task as that of building up the membership and then to fight, by strike action if necessary, for better conditions (2,000 of the 4,000 miners were recruited within twelve months).[6] Davidson's militant attitude, coupled to his personal associations with W. P. Roberts, placed him firmly in the anti-Alexander MacDonald camp in the internal union disputes. St Helens, in fact, was one of the strongholds of the Towers faction. It was J. Atherton, of St Helens, who proposed Towers for president of the Miners' National Association at the first Leeds conference. It was Davidson who proposed that W. P. Roberts should be appointed as the union's legal officer at the second Leeds conference.

W. P. Roberts was a regular visitor to St Helens, either to address meetings or defend miners in court cases. His blunt, outspoken manner made him popular with the men as much as it was disliked by the mine-owners. On one occasion, when he was defending some St Helens miners at the Liverpool assizes for leaving work without authorization, the judge said he had never heard such remarks more calculated 'to embitter relations between master and servant than the speech by Mr. Roberts'.[7]

From the union's inception, industrial relations in St Helens were troubled. In 1863, the men unsuccessfully sued Messrs Johnson, owners of Laffock and Sons and other collieries at Parr, for back pay, illegal dismissal and retention of tools. By February 1864 colliers were campaigning for a shorter working day. They decided to stop work once they had completed eight hours. But the underlooker at Messrs Evans collieries forbade men to ascend the shaft. When they disobeyed, he ordered that water be thrown down upon them. At another pit, a cage was hoisted into a current of hot air and suspended there.[8] The result was that 1,500 men were soon on strike and a demand for a 10 per cent pay rise was linked to the issue of hours.

In the context of the internal union struggle, the St Helens dispute gave the MacDonaldites an opportunity to weaken their opponents. All assistance was denied from the national funds, although Davidson had gained the impression it would be forthcoming. Then Pickard went over to St Helens in an endeavour to break the strike. Coalowners, who had been reluctant to recognize the union or negotiate with Davidson, were nevertheless prepared to see Pickard. At Messrs Evans and Co., he asked the owner, 'Will you yield half an hour or anything to show signs of giving in?' In reply he was asked, 'Mr Pickard, if you were in my place, would you do so?' To which Pickard answered, 'No,' and Evans said that was his answer, too. Whereupon Pickard 'went out of our office', said Evans, 'and tried to induce them to go back to work. I believe that they abused him very much and said that he had been bribed by me.'[9]

Notwithstanding the obloquy, the tactic of Pickard proved to be prudent. The hostility of the employers was too great, and the internal resources of a single district union too small, to sustain the type of militant policy that Davidson envisaged. Soon the strike was smashed, the disgruntled men back at work and George Pickard, William's brother, installed as agent in Davidson's place. On 25 July, 1864, William Pickard told a meeting of 1,500 in St Helens: 'The last time he saw them they charged him with having sold them, when he arranged matters with a Mr Evans, but they found his advice to be good. Davidson had given the wrong advice.'[10]

A change in direction had been achieved at a cost: union membership dropped from 2,000 to 600.

The problem in Farnworth and Kearsley arose from the inability of the firm of Andrew Knowles and Son to reconcile itself to the existence of any combinations among its workers. Employing a total of 3,000 men, it was (to use Halliday's expression) 'the Goliath of the Manchester coalfield'. Within a month of the union's formation, Knowles and Son had sacked between 300 and 350 of its members at the Kearsley, Clifton and Stoneclough collieries. It redeployed labour, bringing colliers from other pits that were not involved in the dispute to those that were, as well as recruiting strikebreakers to fill vacancies. Feelings ran high when the men who had been locked out saw their jobs being taken.

The union realized it was crucial for it not to capitulate before the onslaught of Knowles and Son. At a Wigan miners' rally, in July 1864, Halliday made an impassioned appeal for financial support for the locked-out men. By then the dispute had lasted forty-three weeks and, throughout that time, the union had continued to pay its members

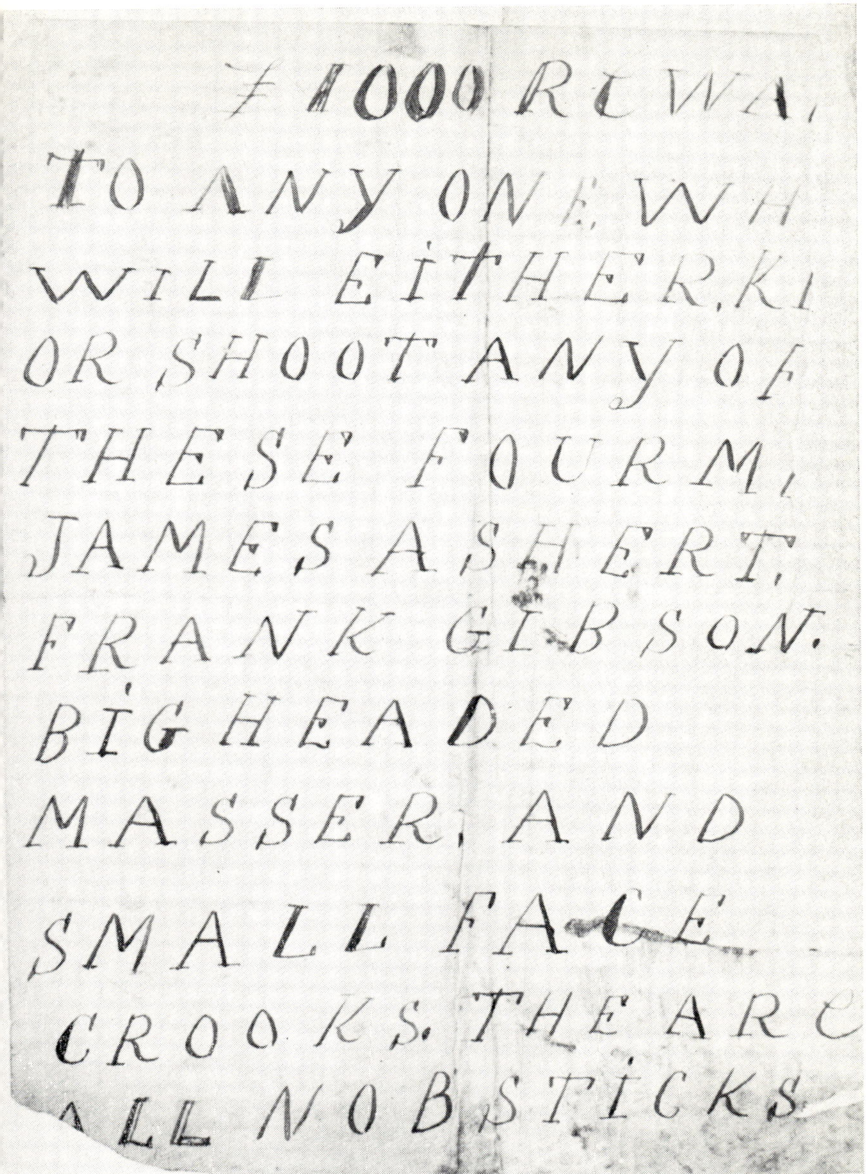
£1000 REWA
TO ANY ONE WH
WILL EITHER KI
OR SHOOT ANY OF
THESE FOUR M
JAMES ASHERT,
FRANK GIBSON.
BIG HEADED
MASSER, AND
SMALL FACE
CROOKS. THE ARE
ALL NOBSTICKS

A warning to strikebreakers, known in Lancashire as 'nobsticks'. This particular note was brought to the attention of the Royal Commission on Trade Unions (1867-9).

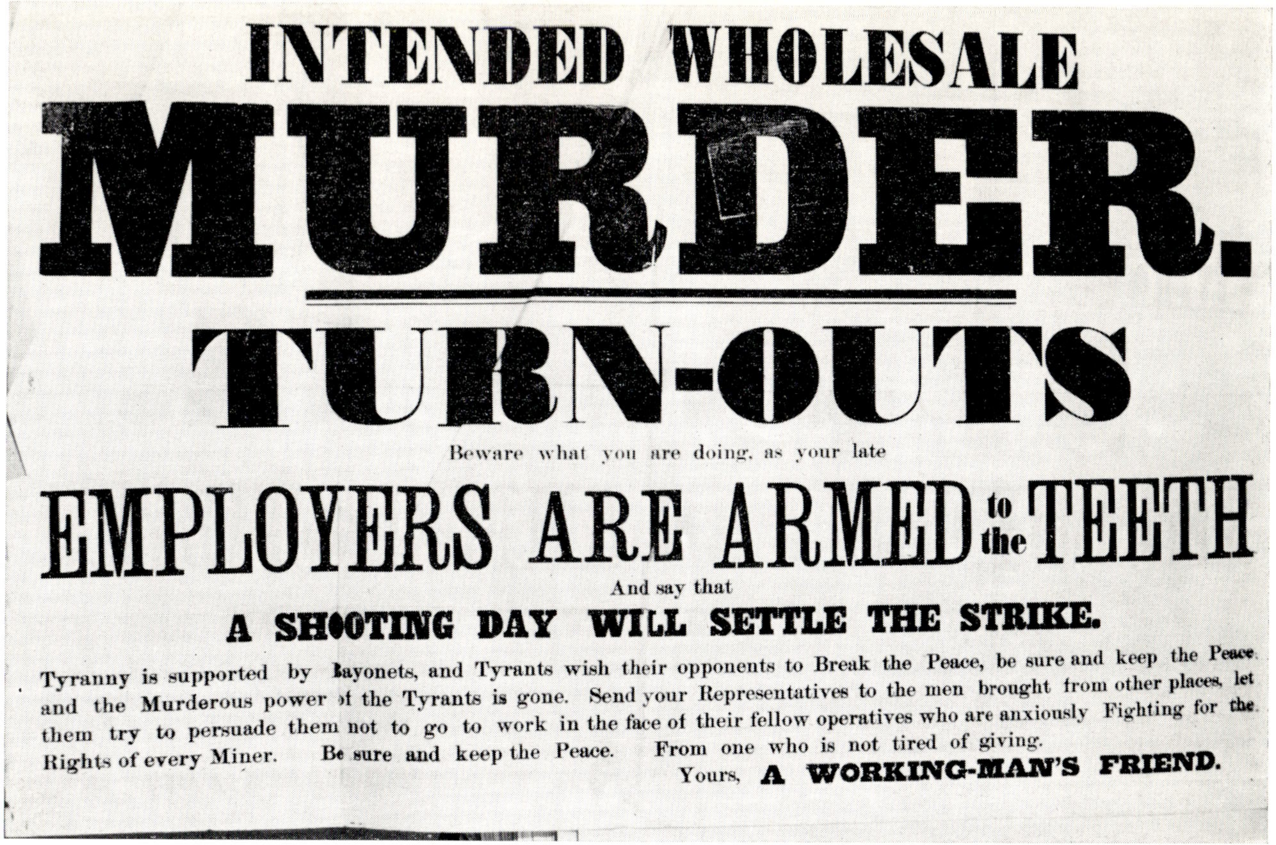

INTENDED WHOLESALE

MURDER.

TURN-OUTS

Beware what you are doing, as your late

EMPLOYERS ARE ARMED to the TEETH

And say that

A SHOOTING DAY WILL SETTLE THE STRIKE.

Tyranny is supported by Bayonets, and Tyrants wish their opponents to Break the Peace, be sure and keep the Peace, and the Murderous power of the Tyrants is gone. Send your Representatives to the men brought from other places, let them try to persuade them not to go to work in the face of their fellow operatives who are anxiously Fighting for the Rights of every Miner. Be sure and keep the Peace. From one who is not tired of giving.

Yours, **A WORKING-MAN'S FRIEND.**

A poster issued during the 1868 dispute. It may have been written by William nine shillings a week, plus a shilling for every child.[11] This was accomplished by imposing a levy on members still working as well as receiving support from other unions. When in its thirty-eighth week, the executive of the Miners' National Association had agreed, rather tardily, to give financial assistance. Halliday reported to the miners' national conference in Wigan, in January 1865, that the dispute had entered its seventeenth month and men were starting to drift back to work. They were signing an undertaking never to belong to a trade union again.[12]

But this was only round one of the contest between Knowles and the union. Some men must have signed Knowles' pledge without the least intention of keeping it. Others probably sincerely intended to keep their promise but, back in their old industrial environment, subjected to the same pressures as before, their response was the same—they joined (or rather re-joined) the union.

While the lock-out had cost the union £5,000, it still emerged from the dispute with £1,000 in the bank.[13] Moreover, surreptitiously miners at many of Knowles' collieries were being persuaded to join the union. Halliday clearly understood the battle was not over.

Lancashire provided many of the national leadership of the Miners' National Association. William Pickard chaired the inaugural Leeds conference. Afterwards he became the vice-president, and subsequently the treasurer.

Pickard was born at Burnley, on 10 February, 1805. Made fatherless at the age of three, he was the eldest child and, as soon as possible, he was sent out to work to help support his mother and the two younger children. When he married, he could neither read nor write. His wife, a pit girl, continued to labour underground until it was forbidden by the 1842 Act.

Keir Hardie described Pickard as 'the very *beau ideal* of the old English country squire of song and story. A real John Bull cast of countenance without the ferocity so often attached thereto, a large fund of commonsense, a fair delivery in speaking and a considerable spice of humour.'[14] Being largely self-taught, Pickard valued formal education highly. His philosophy of life was epitomized by his statement that he

> 'would rather see the miners possessing a higher degree of intellectuality than even an advance of 20 per cent upon their wages, because a man's happiness did not consist of pounds, shillings and pence, but in the honourable performance of duties that were calculated to benefit mankind.'[15]

For a while Pickard contended that because of their lack of knowledge 'the working classes were not qualified to vote and, if they had the franchise, the country would be in a far worse condition than now'.[16] He later revised this view, and unsuccessfully stood for Wigan in the 1874 general election.

Pickard was essentially a moderate, an ardent believer in industrial conciliation. He declared, 'It had ever been his policy—as it ever would be—to try and bring out a unity of spirit between capital and labour. That had been his object from his first setting out.'[17] Inevitably, this brought him into conflict with the militants. They accused him of 'overbearing conduct', during the MacDonald-Towers struggle that caused disruption in the ranks of the Wigan miners.[18] A lay preacher, he was also in trouble with his members for advocating temperance and thrift from union platforms.[19]

When Pickard was appointed treasurer of the Miners' National Association, Thomas Halliday was elected as vice-president. Both men had similar personal backgrounds. When Halliday was two and a half years old, his father died in a colliery explosion, leaving his mother with two children to rear. She went to work in a textile mill. At the age of eight, Thomas became employed at the same colliery in which his father lost his life, and he, too, almost died. In a pen portrait, *The Beehive* described how the young Halliday, 'ascending the shaft one evening, tired by the day's work, in stepping as he thought on the landing, but what was in reality only a strong shadow thrown by the lights carried in advance of him, he fell to the bottom of the ladder, and was picked up as dead'.[20] After this chastening experience, Halliday worked for a while as a 'half-timer' in the warehouse of a spinning factory. Then, his mother remarried. Under the guidance of his step-father, who was also a collier, he returned to the mine, working there a total of 18 or 19 years.[21]

Like Pickard, he addressed temperance meetings, and appears to have encountered the same type of criticism. On one occasion, he was constrained to reply that the miners' leaders were not 'a lot of broken-down teetotallers and Methodists'.[22] He differed from Pickard in being more militant. He strongly advocated the reorganization of the union, a greater degree of amalgamation and centralization, which was ultimately achieved in 1869. At the height of his career as a union leader, the *Bristol Times* described him as, 'a spare, active man, hardly arrived at the prime of life, just the sort of man to do an immense amount of work. He always seems busy, but never flurried, paying great attention to the business at hand'.[23] By dint of his own efforts, he learn to utilize his exceptional talents to the fullest extent.

When Halliday was elected vice-president and Pickard was treasurer, a third Lancastrian—John Worrall, of Kearsley—held another key post, that of secretary of the Miners' National Association. The Rev. Joseph Rayner Stephens, of Ashton-under-Lyne, was appointed as the union's chaplain by the 1863 Leeds conference. So, paradoxically, while Lancashire men had immense difficulty in creating a visible county organization, they dominated the union at national level.

Under the circumstances, it is hardly surprising that Lancashire's voice was heard loud and clear at the Association's half-yearly conferences. One of the main functions of these gatherings was to expatiate on miners' grievances, thereby hoping to win sympathetic support from Parliament and the public. Since this was, by its very nature, a long and arduous process, the same topics tended to arise at conference after conference.

Instead of tediously going through the resolutions of each conference, it is probably better to enumerate the main issues about which miners, especially Lancashire miners, expressed concern:

Long hours of work. In Lancashire the hours of work were among the longest for any coalfield. In Manchester, Bolton and Ashton-under-Lyne, men usually went to work at six o'clock in the morning and did not come out until between half past five and six o'clock at night. At some pits in Pendlebury, men were known to remain down until seven or eight o'clock in the evening. At Wigan and St Helens, there were great variations, but generally men started ascending after five o'clock.[24]

At the first Leeds conference, Halliday moved a solution that eight hours should constitute the working day. He said that excessive labour 'so deteriorates the physical, moral and intellectual condition of the workman as to render it impossible for him to discharge his duties as a man and a citizen'.[25] Likewise Pickard considered the collier's prolonged 'toil left him prostrate and paralysed his mental energies'.

The Employment of Boys under 12 years of age. The union argued that this should be stopped since the heavy work was highly taxing for young boys. Frequently they were so fatigued that, on returning home, they slumped in a chair and had not got the strength to wash themselves and go to bed. Halliday said that tiredness stopped boys from using educational facilities, even when they were close at hand.[26]

In his testimony before the 1866 Select Committee on Mines, Dickinson argued against the union's case: first, that many families could not afford to keep their children at school until they were 12 years old; second, that, as mining was a dangerous occupation, the sooner it was learnt the better; and, third, that the prohibition would have a deleterious effect on the industry since some seams required the use of child labour. Coalowners generally agreed with the mine inspector's conclusions.

Educational provisions. The union wanted these improved. It was necessary (i) to dispel ignorance and squalor prevalent in some mining areas; (ii) to improve standards of safety; and (iii) to create industrial harmony since strife, in the opinion of some miners, arose through lack of knowledge. Obviously, the demand for better education was closely linked to the call for shorter hours.

Mines Inspectors. The union considered that there were an inadequate number of inspectors. There were only two for the whole of Lancashire, Cheshire, North Staffordshire and North Wales. There were, complained Pickard, 'sufficient to run after, but not to go before, accidents'.[27] Another grievance was the inspectors' procedure of informing coalowners in advance of an intended visit, thus giving them time to remedy or conceal any infringements. One miner suggested that, if the police adopted the same method, no criminals would ever be caught.

The weighing of coal. The Mines Regulation Act (1860) permitted owners to continue with the payment of colliers by measure, and this was done in most Lancashire collieries.* But even where the weighing of coal was introduced, with the additional safeguard for the men of a checkweighman, the owners still had much room for manoeuvre. Whether the coal was large or small, good or poor quality, contained dirt or did not were all highly contentious points, factors that could have a considerable influence on a man's wages. An Oldham miners' demonstration was told: 'Although an inspector of weights protected them against Robbery by the grocer, butcher or draper, Mr. Dickinson would give them no protection against robbery of the coalowners.'[28] Continually, throughout the 1860s, Lancashire miners complained about being cheated by the masters. As a man from Ince put it, 'The poor collier has no chance with the masters; for if a tub comes up short a little, they take it; and if one has a little dirt in it, they take it'.[29]

The law and its applications. From experience, the union discovered

* See evidence to Select Committee on Mines, 1866, at the end of the chapter.

the truth of W. P. Roberts' remark that injustice was likely to result from coalowners also being magistrates. When kinship or friendship united the court with the coalowners, then miners thought that partiality would enter into the making of many judicial decisions. They claimed the law was, in any case, biassed against the men. Miners could wake early in the morning, walk three or four miles to work for 6 a.m., only to discover the colliery was not going to open that day. While they had no redress if the owner made work unavailable, they were liable to one months' imprisonment under the Master and Servant Act, should they absent themselves. This Act, a Wigan rally was told, was 'a relic of a barbarous age' and created a grossly unjust position: 'If John the miner was liable to be dragged from his bed by the neck for breaking his contract, John the master must be liable, too, or the laws of England were unequal.'[30]

The payment of wages. The union called for wages to be paid weekly. The St Helens colliers thought it should be on a Friday, thereby enabling them to do shopping on Saturdays. But the coal-owners argued this was impractical. In Lancashire payment was usually made fortnightly on 'rechinin' Saturday'. Since the miners did not plan their expenditure very well, there was a fortnight cycle with the transition from affluence to penury. It began with three-day bacchanals, beer-drinking and eating:

> 'Supremely wasteful is then the feast. Butter and unctuous things (for they like it rich) make an ocean for cakes and meats and good things to bathe in. Fires, never small, heat their rooms, while the fat streams out of the ovens, and they keep it cooking away till far into Sunday morning. Look in upon them, and in two minutes you feel ready for basting, and make a quick exit, for fear of being served up with the roast meats. Drop in next Saturday, you are in danger of a different basting.'[31]

The provision of supplies, powder and candles. The union claimed that where miners were expected to buy these things from company stores, customarily the prices were high and the coalowner's profits excessive. The miners fought for the right, and in many places gained it, to make their own purchases, independent from the owner. But this course was fraught with danger. It meant the storage of gunpowder in numerous cottages. Sometimes, explosions destroyed a house—indeed, in one instance a whole row—and loss of life ensued. The disregard for elementary safety is illustrated by the fate of an Aspull man. He was carrying 125 lbs of gunpowder home from Wigan, having drinks at public houses *en route*, and then, at New Springs, somehow his cart tippled over, spilling the powder. Assisted by other people—and smoking his pipe—he started putting the gunpowder back into the cart. The *Wigan Observer*, commenting on the incident, said:

> 'The moral of the story is that to collect gunpowder when smoking a pipe is likely to result in a serious accident, and the only extraordinary thing about the business is that the deaths of two persons have been required to prove the fact conclusively to a number of thoughtless Aspull colliers.'[32]

The Explosive Act 1875, imposed stringent restrictions on the sale and storage of gunpowder. But even this encountered opposition. In 1877, the South Lancashire and Cheshire Coal Association's annual meeting heard Major Ford, a government inspector, ask for coalowners' assistance in eradicating the habit, prevalent in the Lancashire coalfields, of miners manufacturing their own blasting cartridges. They had to be made, he said, in a duly authorized factory.[33]

The employment of female labour in pitbanks. The union declared to the Select Committee on Mines, 1866, that this practice was 'a foul blot on the civilization and humanity of the United Kingdom'. In fact, it was a practice almost entirely confined to the Lancashire coalfield and, even there, it was limited to certain areas. The main centre appears to have been Wigan, where about five hundred women were employed, although the numbers were slightly swollen by the Cotton famine.

The type of work they did was realistically described by John Monk Forster, a Wigan collier who became a writer, in his novel, *The Pit Brow Lassie*.

> 'No work that woman or girl could do would be found in the hands of boy or man. There was a woman helping the banksman at the cages; women and girls were running full and empty tubs in all directions across the pit brow which, like the sides of a modern warship, was covered with smooth, thick plates of iron over which the small wagons could easily slide; women and girls were busy in the shoots and screens, freeing the coal from dirt and slack. At the end of the great coal stack, where coals were stored when trade was bad, a group of women, with only here and there a man among them,

were busy with spades and riddles filling railway wagons, and each woman handling her implement of labour as easily and deftly as a man. On the other side of the pit brow, where the Leeds and Liverpool canal ran, another group of women were filling a boat with coal in the colliery basin under the guidance of a man.'[34]

The women, who were aged between 12 and 50, earned from 1s. 6d. to 2s. 6d. a day. Being common labourers, they dressed for the part in clothes frequently the same as the men:

'Clad in coarse, greasy and patched fustian unmentionables and jacket, thick canvass short, great heavy hobnail boots, her features completely begrimed with coal-dust, her hard and horny hands carrying the spade, pick, drinking tin, sieve and other paraphernalia of her trade.'[35]

Emily Faithfull, who gave this description to the Social Science Edinburgh conference in 1864, said many of the women smoked pipes, drank inordinate quantities of alcohol, and one had gained 'a certain notoriety by her pugilistic exploits'. Another report spoke of the difficulty of distinguishing men from women, the main difference being that women pit brow workers sometimes wore bonnets.[36]

Most outside observers regarded the use of female labour in this way disgusting and degrading, a slur upon womanhood. But what makes a person repulsive to somebody who is not a Lancashire miner may be precisely those features that add enchantment to one who is. That beauty is evidently in the eye of the beholder is clearly shown by the poem, 'A Pit Brow Wench For Me':

'I am an Aspull collier, I like a bit of fun,
To have a go at football or in the sports I run;
So good-bye old companions, adieu to jollity,
For I have found a sweetheart, and she's all the
world to me.
Could you but see my Nancy, among the tubs of coal,
In tucked up skirt and breeches, she looks exceeding
droll.
Her face besmear'd with coal-dust, as black as black can be,
She is a pit brow lassie, but she's all the world to me.'[37]

According to William Pickard, ninety-nine out of every hundred pit brow lassies ultimately married colliers.[38]

Besides arguing that heavy manual labour was not the kind of work appropriate for women, the union claimed it had a harmful effect on family and community life. Coming home tired from the colliery, women were unable to spend the same time keeping their homes clean and tidy. The children also suffered. They lacked a mother's guidance and control. Women working in a rough, tough environment were likely to acquire uncouth habits which they passed on to their children.

Replying to these criticisms, coalowners claimed that the union did not speak for the pit brow women themselves, who enjoyed their work. It would be an infringement of their liberty to stop them from working. What is more, it may be, extremely difficult to find alternative employment, which anyway might be more tiring and exacting. Mr Burrows, the Atherton coalowner, declared that pit brow lassies were 'robust and well developed in physique, look as healthy and as happy as any class of labouring women I have ever met; and not, so far as I could see, so hard worked as many a toiler at a sewing machine'.[39]

Lurking behind both sides' concern for the women's well-being stood their own respective economic self-interest. J. M. Forster went so far as to claim that 'the whole movement against pit brow women had its origin in the worst of motives. No feeling of charity actuated the men who took up the cry; they had no chivalrous regard for the weaker sex; it was not because they deemed the labour too ardous or that it had a tendency to demoralise the worker; it was because they regarded bit brow women as rivals in the labour market and wished to have the whole field to themselves.'[40] Similarly, Emily Faithfull showed that the coalowners had a hidden economic motivation: 'The truth is female labour is plentiful and cheaper, far cheaper, than that of the men ... employers will be found very slow in encouraging a change which may diminish their profits.'[41]

Towards their grievances, miners were impelled to adopt a stoical attitude. Most of them had been endured for a long time and, since they required legislation there appeared no immediate prospect of improvement. Miners might grumble and complain, yet there was little they could do other than show disapproval.

In another sphere—that of union organization and men's wages—the miners' response was entirely different. Any attempt to reduce pay or smash their organization was likely to arouse an immediate and violent reaction. It threatened their whole independent economic existence.

This is precisely what happened when Knowles decided to make a determined effort to smash trade unionism in his collieries. He had always told his managers and overlookers to maintain a strict surveillance, dismissing any men suspected of union activity. But in August 1866, when newspapers reported that 1,600 of the 6,000 colliers attending a rally at Hollingworth Lake came from Farnworth and Kearsley, he realized how ineffective his measures had been. Apparently, most of the men from Knowles' collieries had completed their shift in the forenoon before going to the rally. This made detection difficult. Eventually, it was resolved to ask each man individually, 'Art thou in the union?' Those who said 'Yes,' or refused to answer, were immediately sacked. As a result, all the men at Fearneysides pit, Little Lever, were dismissed. At Bank Top colliery, Bury, 170 men were discharged. The same thing occurred at Pendlebury and Pendleton until five hundred of the 1,400 miners who worked in these particular pits had been dismissed.[42]

The union's first reaction was to try and avoid a struggle. Halliday went to the offices of Andrew Knowles and Son, King Street, Manchester, in an attempt to settle the matter by negotiation. Knowles refused to see him.[43] Then, on 12 December, 1866, the committee of the Pendlebury colliers respectfully asked Knowles why he had taken his decision and pleaded with him to meet a deputation.[44] They received no reply.

With no prospect of conciliation, the union resolved to fight. Men declared themselves to be union members and left the pits in solidarity with their sacked brothers. By the beginning of 1867, the Farnworth and Kearsley district had 1,447 of its members locked out. It paid them nine shillings a week, plus a shilling for each of their 1,266 children under 12 years.[45]

To counter the union's resistance, Knowles decided to recruit large numbers of strikebreakers. Three thousand notices were printed. Paul Roper was sent to Staffordshire. He received a pound for every man he engaged, as did Colonel Goodier, who was involved in recruiting scab labour from Cornwall. As a result, the firm acquired the services of five hundred Lancashire men, three hundred from Staffordshire and two hundred from Cornwall.

The first consignment arrived from Staffordshire in two horse-drawn buses on 15 December, 1866, and Halliday advised the men to take the threat seriously.[46] Consequently, the union dispatched two of its representatives to South Staffordshire in an endeavour to dissuade colliers from coming. They appear to have had only limited success.

The main problem facing Knowles does not seem to have been the recruitment of strikebreakers but finding them accommodation. In part, he solved this difficulty by evicting families from tied cottages belonging to the company. Many were on short tenancies agreements, where a week's notice had to be given. In this way 27 families were removed from their homes in Clifton, principally in Moss Colliery Row and Turton Row, while probably about three times that number were similarly treated in Pendlebury and Little Lever.[47]

Even so, accommodation was too limited. So Knowles converted many single dwellings into double dwellings. The door between the living-room and the kitchen was blocked in, a further staircase added, and two self-contained housing units had been created.[48] It was a common sight to see two strikebreakers, returning home from work, with one going in at the front door of a cottage while the other entered the back.

Another stratagem used by Knowles was to find lodgings for his new employees. Most of the indigenous inhabitants, loyal to the strikers or frightened of openly not being, refused his blandishments. So the residue of Staffordshire and Cornishmen had to be housed in store-rooms and specially erected, or converted, sheds.

In March 1867, a magistrate, sentencing a striker to six months' hard labour for fighting with a knobstick outside one of these establishments at Stoneclough, commented, 'Upon this building the wroth of the unionists was frequently expended and the whole of the windows have been broken on at least three occasions'.[49] In the same month, a fire broke out in the store-room of Pendlebury colliery, which contained oil, tallow and hemp.[50] Had it not been for prompt counter-action, the quickly-spreading blaze might have reached a nearby shed where the colliery's gunpowder was kept, another shed which housed blacklegs and their families, and a gasometer on the adjacent land. As it was, Knowles and Son sustained a loss of £800—£500 worth of stores and £300 to property. In addition the company's Ringley colliery had its boiler blown up.[51]

Clearly, the strike posed problems for the police. They sought to stop the almost nightly fights between Staffordshire and Lancashire men. Superintendent Bent used his Wedgwood stick to disperse men whenever crowds began to congregate.[52] Four men were even arrested for standing outside their own front doors. W. P. Roberts, defending, declared, 'The magistrates were being used as tools by Messrs Knowles to manipulate the law in their favour ... it was wrong for them to convict the defendants for standing on their own doorstep.'[53] Nevertheless, police vigilance was understandable. A Staffordshire miner had been killed in one fight without the authorities being able to apprehend the assailant. On numerous occasions, they revealed that they were incapable of providing blacklegs with adequate protection, either on the way to work or at home. A typical case was that of a knobstick called Holford, who was sleeping with his wife and child when a bomb was thrown into the bedroom. He was also threatened 'that he would rue the day he ever came to Pendlebury'.[54]

Supporters of the strike seized every opportunity to display their hostility to the interlopers. Crowds of women, banging pans and shouting obscenities, accompanied the police convoys that escorted the blacklegs to work. They ostracized them socially so that a ghetto, known as 'Old Bilston', grew up in the Bridge Street area. Any Lancashire lassie who married 'a Staffordshire stallion' had to sever her links with her old associates and join this closed community of immigrants. According to a local paper, almost 100 years later inhabitants of the Bridge Street area still possessed some of the customs and dialects of their ancestors from the Black Country.[55]

All these moves, proved to be in vain. By May 1867, the union conceded defeat in an extremely abject way, as this letter to Knowles and Son from the Pendlebury lodge shows:

> 'Sir,
>
> We are ashamed of writing to you through our conduct in the strike, knowing that we put you to a great deal of trouble and expense by our unmanly conduct, but we were led on by men who said fine things and made great promises (who, we believe, had had their feelings hurt because you would not speak to them). Fools that we were to heed them, men who we believe are very little for us so long as they can get their 30s. per week and themselves thought to be somebody. But we have had enough, and more than enough, and many of us were young and inexperienced in such things, but thought at the time it was wrong to strike in the manner we did without a moment's notice but we were over ruled; fools I say we were again to heed them.
>
> But now to the point. We humbly beg you that you will look over our conduct for the past, and we promise amends for the future. We are forced to keep in the union while we are out of employment; you will see that, if we withdraw from the union, and then you not employ us, our support would be gone, and we should have to leave the country to seek other employment. We were led to join the union under the belief that it would do us good, and the rules went to say that we were not to do anything that would injure either employer or employed, rules that have been foully broken by the present strike. But we, the leading members of the Lodge

which is held at the Windmill Inn, and which is the most influential lodge in the whole district, pledge ourselves to withdraw from the union if you will overlook our fault this time, and take the few members we have out of employment to their work again, with the privilege of turning our lodge, as we have some money on hand which we do not want to leave, and we also think that if you will grant us these things, that turning of our lodge into a sick lodge will be the means of smashing up the union in Pendlebury. Hoping you will consider our petition, and if you can grant us our request and we promise not to be so soon led to enter a trades union again.

Signed on behalf of the leading members of the above lodge,

I remain, Your humble servant,

JOHN CHAPMAN

Secretary.

P.S. You must not think that it is on my own account that I write this. I was not working for you at the time of the strike, and I do not think I shall ever work in the pit again.'

A week later, on 23 May, 1867, came the reply to this letter:

'Sir,

In your letter of the 16th Instant, I can enter into NO negotiations on the subject therein named.

Yours truly,

JOHN KNOWLES'[56]

Knowles won the struggle because the size of his industrial empire gave him a strength a local union could not match. At the height of the dispute, the union were only able to call 1,500 men out. This meant that Knowles and Son had half its labour force unaffected. Many collieries could still continue—indeed, expand—production to compensate for those involved in the dispute. When the company decided to supplement this tactic by the mass importation of strike breakers, the ultimate outcome was not in doubt.

For the union, defeat was a severe blow. Halliday admitted 1,000 members had been lost and four of the 52 lodges closed down.[57] The six-month struggle, costing £16,000, had depleted union funds. While this appears to have been the initial impact of the defeat, when the full consequences became clearer to the men in the form of lower wages, poorer conditions and unemployment, it seems that demoralization was much more widespread. Many of the men on strike had failed to get their jobs back; those members at work during the lock-out paying a five shilling a week levy, appeared to have nothing to show for their sacrifices. It seems probable that the membership of the Farnworth and Kearsley district, which was five thousand in 1866, had slumped to a little over one thousand by the end of 1868. A district union lay ruined.

The same pattern was repeated in Wigan. A recession in the coal trade had resulted in stockpiling and the owners demanded a fifteen per cent wage reduction in February 1868. Immediately this was announced William Pickard, who was usually in contact with Alexander MacDonald two or three times a week, asked him for his advice. MacDonald replied it would be 'disastrous and foolish to attempt resistance'.[58] Both Pickard and MacDonald attended the Wigan delegate meeting on 21 February, in an endeavour to dissuade men from withdrawing their labour. MacDonald described what happened: 'When there I found that ever delegate had come with a written order from his men "Strike".' He asked them to re-consider their position, and the delegates agreed to consult their members again. Meanwhile, MacDonald and Pickard sought to influence the men directly, over the delegates' heads, by holding a public meeting. Ten to twelve thousand heard them suggest a compromise formula: to tell the owners they were prepared to forego the last increase granted, usually ten per cent. But the miners remained determined not to budge. When the second delegate meeting was held, it was resolved to go on strike. As MacDonald made clear to the Royal Commission, 'It was not the leaders of the union they consulted.'[59] He went on to say that both Pickard and himself disapproved of the strike.

Nevertheless, among rank-and-file miners it gained mass support. By 13 March the strike had spread throughout South Lancashire. In Wigan ten of the thirteen thousand colliers were out and in St Helens it was six thousand. Most of those remaining at work had only received small cuts. A newspaper reported 'union and non-union men have made common cause'.[60] Non-union men, who received no financial assistance from union funds, had a public collection started for them.

Many meetings were held. At one in Wigan, addressed by Pickard and MacDonald, another attempt was made to sell the idea of accepting a ten per cent cut as a compromise. When put to the meeting, there was a confused reply. MacDonald's speech was noteworthy because, although the colliers were in the middle of a strike, he never mentioned it once. He confined himself to talking about the need for better education, greater safety measures and weighing machines. Pickard, who told the meeting that they were on their way to give evidence to the Royal Commission on Trade Unions, expressed concern lest they did not remain on their best conduct while he was away.[61]

Pickard's fear proved to be well founded. A few days later, seven hundred men, armed with picks and pickshafts, made a midnight attack on Richard Evans and Co., collieries at Haydock. Soldiers were quickly dispatched from Warrington. J. Evans, the proprietor, told the Royal Commission on T.U.s what happened after he had addressed the crowd, saying that there were ample police and soldiers to cope with any emergency:

> 'Before I could turn my back, they called out, "Give it them", and a shower of paving stones were thrown among the police, immediately after the police charged them, and sent them to right and left across the fields. There were a great number of men injured, five or six of the police were invalided, and one superintendent had a most serious wound in his face.'[62]

This was the first blow in a wave of violence that spread across the coalfield. Threats of violence to individuals, such as the notice (see illustration): '£1,000 reward to anyone who will either shoot or kill any of these four men—James Ashert, Frank Gibson, Big Headed Masser or Small-face Crooks.' William Rowe, a strikebreaker at Lindsay pit, who received a similar warning, told the Royal Commission it was no idle threat since the police usually 'came after the thing was gone over'.[63] Another Lindsay blackleg, J. Johnson, shuttered the inside of the windows of his house at Whelley. He left his wife and eldest son on guard downstairs while he went to sleep with his other ten children. At midnight shots were fired into the house, and at two a.m. a gingerbeer bottle, filled with gunpowder and lighted through a wooden plug, was flung in.[64] According to one of the coalowner's chief spokesmen, there was more violence than in any previous strike.[65]

But the main acts of violence were made by large groups of men. The usual tactic was for men from one district to come to a district where they were unknown to the authorities. Often they wore black crepe hoods as a further form of concealment. Five thousand colliers from St Helens, Haydock and Ashton marched northwards to Wigan damaging property at any colliery that continued production. The most grave incidents happened at the collieries of Wigan Coal and Iron, where the management had decided to employ Staffordshire strikebreakers. More than five hundred attacked the Gidlow pit, trying to break the winding-gear and setting fire to the engine house. Alfred Hewlett, the managing director, said that the strikers then stood around, watching the blaze, 'and used expressions I would not like to repeat'.[66] At Lindsay pit, one thousand five hundred colliers drove the Staffordshire men out of the colliery and then took possession. Only the arrival of the military succeeded in dislodging them.[67]

Against the use of force by the strikers on such a scale, the authorities admitted their incapacity to cope with the situation satisfactorily. Colonel Bruce, chief constable of Lancashire, stated that he only had eight hundred men under his command for the whole county; they were insufficient to protect property. In Wigan and the outlying districts there were ninety policemen and the magistrates, recognizing that they would be unable to protect the Staffordshire strikebreakers, actually ordered the Wigan Coal and Iron Company to send them back home.[68] At five a.m. one morning, the blacklegs departed to the accompaniment of 'loud cheers mingled with hisses and groans' from the triumphant local populace.[69]

The violence of the Lancashire colliers aroused widespread dismay. The Earl of Lichfield, speaking in the House of Lords on 1 May, 1868, said there were grave dangers implicit in the victory of the Wigan mob. Cavalry should continually ride through coal mining districts as a warning. In the House of Commons Major Anson expressed the fears that elsewhere 'miners may be encouraged to pursue a similar course in expectation of similar results'. The *Colliery Guardian* almost certainly voiced the opinion of the majority of property-owners about events in Wigan when it said, 'The security of every interest worth preserving in the nation has been put in jeopardy.' For *The Spectator* the answer to the colliers lay in draconian repression, 'Five hundred lives ought to have been taken in Wigan rather than those five hundred labourers should have been robbed by violence and impunity of their labour.'[70]

In its early stages the strike had led to many court cases. Alfred Hewlett, of Wigan Coal and Iron, stated that in the first nine weeks he alone was responsible for taking 64 cases. But the fines imposed had been ten shillings or a pound, a punishment that had little deterrent value. However, as the strike continued, courts started to adopt a tougher line. Police armed with cutlasses guarded the Quarter Sessions, where the Recorder of Wigan condemned the strikers because they had 'instituted a system of terrorism which, if it had not been immediately suppressed, must be subversive to all law'.[71]

The harsh sentences meted out by the Recorder created, among many colliers, a lasting bitterness. Twelve years later, in the general election of 1880, a leaflet recalled how John Lancaster, a local coal-owner who was also M.P. for Wigan, had refused to intercede on behalf of the imprisoned men:

> 'Have you forgotten the 1868 Strike, when Samuel Lancaster sent for 300 Staffordshire Men to the Sawmills to work at the Moor Pit, and they brought them down in waggons with planks over them? We had a procession, and William Pickard led us up to the Saw-mills, and when we returned he gave us a speech at Boyswells Fields. In that speech he called Mr. Lancaster whitewashed walls and other things next nearly to being a murderer. After the meeting we returned home comfortably, but in a few days after there were men put in prison and others shot at. The soldiers came to calm things down; but, after all, there were five men put in prison for terms varying from 18 months to 5 years, and one of them died there; and in that struggle Lancaster, the colliers' friend—accompanied by Mr. Pickard, our supposed guardian—made his escape into the Minorca Hotel for safety from the mob that he had caused with a revolver in his hand. When the strike ended, and work began, the parents of the imprisoned men asked Mr. Lancaster if he would do his best to get them out, and he said No.'[72]

The 1868 strike ended raggedly. In the early stages, some of the owners, especially the smaller ones, agreed to the miners returning to work with little or no reduction. By April, agreements reached usually involved at least a ten per cent reduction. But big firms, like Wigan Coal and Iron stuck out for a full fifteen per cent reduction, and they gained a complete capitulation to their terms from a delegate meeting in early May.[73]

For the union, the lesson was the same as that of the Farnworth-Kearsley strike: organizations formed on a district basis autonomous and independent from all others, were no match for the gigantic industrial empires that had emerged. Wigan Coal and Iron Company, for example, employed 9,000 men and had an annual production of $1\frac{3}{4}$ million tons. Pitted against it was the puny Wigan union, whose £10,000 reserves were exhausted after only three weeks struggle.[74]

By 1869, all the district unions in Lancashire had collapsed. They had failed to protect their members from the employers' onslaught that had come with the trade depression. It was clear that a new form of organization was necessary.

Table 7 *Date of Formation of District Unions*

Wigan Miners' Provident Benefit Society	5 September, 1862	(1)
Farnworth & Kearsley Miners' P.B. Society	17 August, 1863	(2)
Worsley Miners' Association	20 August, 1863	(3)
St Helens Miners' Association	3 July, 1863	(4)

Sources: (1) *Wigan Observer*, 12 September, 1862.
(2) *The Beehive*, 17 October, 1866.
(3) *Miners' and Workmen's Advocate*, 2 February, 1864.
(4) *Wigan Observer*, 3 July, 1863.

Membership of District Unions

	Wigan	F & K	Haydock	St Helens	Worsley
28 Nov. 1863	3,000				
22 Oct. 1864	4,000			2,000	
19 Nov. 1864	4,258	2,500	258	600	
25 Nov. 1865	4,368	2,600			600
29 Jul. 1867	6,000	4,000	1,200	1,200	800-900

Ashton-under-Lyne and Oldham Districts never reported to conferences on their membership figures. Nor is it possible to say when they were formed.

The Manchester conference of July 1869 was told that membership in Lancashire and Cheshire had declined to a total of 6,500.

EVIDENCE TO THE SELECT COMMITTEE ON MINES, 1866, FROM LANCASHIRE WITNESSES

Peter Dickinson, of Aspull, a collier at Wigan Coal and Iron Co., said he had never worked in a Lancashire colliery where the men were paid by weight (Q.892). It was a long grumble of the colliers.

Robert Woodward, of Haydock, employed by Messrs J. & T. Johnson, stated that they had applied to the master to be permitted to appoint a checkweighman (Q.2815), but he had refused (Q.2812). 'We suffer greatly from it' (Q.2821).

Joe Booth, a miner of Hyde and Haughton colliery, Cheshire, said that there was general dissatisfaction because they were paid by measure (Q.4588).

William Baxendale, a miner who worked for Robert James & Co., of Ince, referring to the appointment of checkweighmen, said, 'I do not consider that we have an Act that gives us the power because, if the owners put the men off the pit-top after they are appointed, we have no advantage' (Q.1573).

William Pickard, union official, said payment was usually by measurement rather than weight (Q.2382). 'The law is very unsatisfactory at present.'

Maurice Mullin, a checkweighman at Ince Hall Coal & Cannel Co., stated that, before his appointment, the men were forfeiting an average of 99 tubs a day; now they lost only nine (Q.1931).

G. Gilroy, a colliery manager from Wigan, said he did not think the system of payment by weight could be applied in South Lancashire and Cheshire (Q.10876). The coal was frequently wet, and hence the weight did not accurately reflect the amount hewed.

CHAPTER FIVE

Lancashire and the Amalgamated Association

By the beginning of 1869, the Lancashire district unions lay in ruins. They had been unable to resist the effects of trade depression and the concomitant attacks on miners' wages. In April of that year, the employers struck a further blow when they announced there would be another reduction of ten per cent.[1] At Tyldesley, where the men resisted, the pattern of previous struggles was repeated: the temporary closure of the mines, the importation of Staffordshire strikebreakers, the resumption of production, the colliers' capitulation. But on this occasion the defeat was even more ignominious. A week after resuming work on the employers' terms, the men were confronted with the prospect of another wage cut, this time of 12½ per cent.[2] All the gains, accrued since the advent of district unions in 1862, had been eroded. A union leader confessed that, when allowance was made for the higher cost of food and the larger deductions made from wages colliers in Westhoughton were receiving an income, in real terms, that was twenty per cent lower than in 1862.[3] The position varied from place to place. The most favourable estimate came from the Wigan area, where Pickard estimated real wages were still five per cent above the 1862 figure.[4] Nevertheless, this did not alter the overall situation and the lesson to be drawn from it. The disarray of the miners' organizations, their greatly diminished membership, activity and cohesion, testified to their failure to meet the challenge of events.

Nobody realized this more than Thomas Halliday, the architect of the Amalgamated Association of Miners. He saw the district unions were too small, their financial resources too slender, to do battle with the industrial giants that were emerging in the coal industry. The Farnworth and Kearsley union, where he was miners' agent, had been defeated by Andrew Knowles & Son. During the dispute it had been compelled to levy the punitive sum of 2s. 6d. weekly on all its members not involved in the lock-out. As a consequence, many miners, finding the financial burden too heavy, left the union. Throughout the county the whole structure of the unions, their division into independent and rival entities, encouraged mass desertion, the temptation to move from one district, where high levies were being paid, to another, where they were not. In the middle of the contest with Knowles, two lodges deserted Farnworth and Kearsley *en masse* and joined the Worsley Miners' Association, thereby swelling its membership by almost a third. This showed that, at times of greatest need, when unity was most urgently required, there was always the greatest likelihood of division.

But Halliday saw that the flaws in the county organization only served to emphasize equally serious defects in the national organization. When Knowles had tried to wipe out the Farnworth and Kearsley district, and almost succeeded, the Miners' National Association did little to assist its hard-pressed brethren. The sole financial support given was a threepence levy from each member, a totally inadequate amount. Halliday concluded that, instead of the existing loose, federal structure, there was a need for a centralized national leadership. Then, the entire financial resources could be mobilized against any individual or small group, of employers who sought to wreck the union in their locality. Such a tightly-knit organization among the men would tilt the relationship of forces in all coalfields, giving miners greater powers *vis-à-vis* their employers, and making it possible to secure higher wages that could be gained by continuing with the haphazard, unco-ordinated manner that had become traditional among colliers in the 1860s.

These were the underlying reasons for the formation of the Amalgamated Association of Miners.

As the Webbs and others have pointed out, the principle of centralization as generally applied by trade unions in Victorian times was usually associated with a policy of moderation and class conciliation while decentralization connoted a more militant policy.[5] To understand why precisely the opposite applied in mining trade unionism, involves an analysis of its development during the 1860s. It was a time when the most contentious issue, the one that conference after conference debated, was what attitude the national organization should adopt towards industrial disputes.

Gradually, a polarization of opinion, two irreconcilable camps, began to emerge. At the Manchester conference in November 1864, one of the main reasons for the split, the breakaway of the Practical Miners, was the failure of the M.N.A. to support financially or even morally, the struggle of the Staffordshire colliers. When their strike ended in defeat, with the collapse of the Practical Miners as well as the Staffordshire district union, different lessons were drawn. The reaction of Alexander MacDonald was 'to denounce all strikers because they could accomplish all they wanted by legal and constitutional means, and be the better for it'. In contrast, Halliday thought the Midland colliers 'wanted to feel they would not be deserted in their times of need. If the Staffordshire men had been supported by 200,000 men, they would have felt and acted very differently.' The M.N.A.'s Wigan conference (January 1865), where these two viewpoints were expressed, opted for the MacDonald position. It proclaimed that all industrial disputes should be resolved by arbitration and conciliation; financial assistance would only be given in lock-outs. Six months later, at the Tipton conference, the decision to levy the membership to support men locked-out was even revoked. William Pickard argued it was beyond the national association's resources to provide such cover while another of MacDonald's supporters, Dr Miller, representing Silverdale in North Staffordshire, averred that 'the miners felt the utility of some bill to end strikes'.[6] In its report, *The Beehive* stated the Tiptop conferences 'was not occupied with strikes or any matters directly affecting wages'.[7]

The following half-yearly M.N.A. conference, at Newcastle-on-Tyne (December 1865), reversed the trend towards industrial pacifism. Nathaniel Allan, of Worsley, successfully moved the adoption of a new, elaborate procedure. It laid down that whenever a dispute arose, a local union would forward details to the national committee: numbers involved, weekly cost of resistance and what it thought should be done. Then, the national committee would consider it and forward its recommendation, whether to levy the membership or not, to the districts for discussion. In the light of their reactions, the national committee would eventually decide what attitude to adopt.

This cumbersome procedure proved to be unworkable and a source of irritation. The national committee, controlled by the MacDonald faction, stipulated that it must be fully informed and give its approval before a local union went on strike, which was only to be used as a last resort. In most instances, however, where relations between owners and miners deteriorated, an industrial dispute would flare up before the issue had been considered by the national committee. Hence, the local union would be held to be in violation of the rules and not entitled to benefit.

The Oldham district complained about this to the M.N.A.'s Bolton conference (June 1867), only to be told by the president that they should have tried arbitration not striking and that their action was unconstitutional. Endorsing this view, Burt and Normansell moved that the matter be left to the districts, a polite way of deciding to do nothing, for they knew the larger county unions, like Northumberland and Yorkshire which they represented, would make no contribution. At the same conference, the Ashton-under-Lyne district applied for £465 6s. 0d. It had successfully gone on strike to compel the coalowners to concede the eight-hour day. Since this was official union policy, the Ashton delegates thought they should receive union backing.[8] But the moderates argued that as the resolution favouring the eight-hour day contained no specific reference to giving financial support, it was a matter that was entirely up to the local unions. Conference endorsed this line.

Both wings of the M.N.A. discovered existing arrangements to be unsatisfactory. The moderates, controlling the national committee, could usually devise some formula that shook off liability to pay money to men in dispute. Nevertheless, the wording of the Worsley resolution did, at least theoretically, commit the M.N.A. to assist in some circumstances. To leaders like MacDonald, hankering after respectability and wishing not to antagonize their influencial friends in Parliament, even this verbal commitment proved too embarrassing. After the Sheffield outrages, when gunpowder was put down the chimneys of houses where blacklegs lived, followed by the public outcry over the violent struggles of the Lancashire miners (1867-8), MacDonald thought it imperative to get the M.N.A. completely disassociated from all forms of industrial strife. Early in 1868, the M.N.A. rescinded the Worsley resolution; it would not even support men locked-out.

This must have been very galling for Halliday. Events were pushing him in the opposite direction. The Staffordshire strikebreakers who had come to Farnworth and Kearsley, smashing the strike, had vociferously complained about their treatment during the Staffordshire dispute of 1864-5. Lancashire colliers and others had given no financial support; the only thing that had come from Lancashire was coal that had helped to defeat them. So they felt quite justified—it was a form of retribution—to work for Andrew Knowles & Co.

Internecine strife, coupled with a lowering of wage rates, appeared to be what the future held for miners unless they could establish some new form of organization. This was how Halliday saw things in 1868.

The contrasting outlooks of Halliday and MacDonald were clearly illustrated in their testimonies to the Royal Commission on Trade Unions in 1867.[8] The Commission received MacDonald's evidence in a cordial, friendly atmosphere. They listened sympathetically to his denunciation of strikes and how the M.N.A. now no longer even supported men locked-out. They heard him suggest that coalowners should take intelligent colliers into their confidence, inform them in advance of wage cuts, so that they could help persuade their fellow miners to be more amenable. This, said MacDonald, was better than baldly putting up notices of wage reductions, a course more likely to create conflict.

While the Commissioners allowed MacDonald to reflect on his undergraduate days at Glasgow University, his appreciation for the professor of Classics, they gave Halliday a much rougher passage. In their eyes, he was the leader of the Lancashire strikers, a man who was probably implicated in the various acts of violence. They regarded him as a hostile witness from whom they wanted incriminating evidence. And it must be added that some of his replies did not appear particularly convincing. For instance, referring to the fire at the stores to Pendlebury colliery, he was asked, 'You heard that it was attributed to the union?' To which Halliday answered:

> 'Speaking truthfully, I had not heard before this day that it had been blamed on the miners' union. It did occur to me that another thing might have been attributed to us, and that was a boiler that unfortunately exploded at Ringley. I was sorry for that because it was a destruction of property.'[9]

The two differing conceptions of the way forward gained practical expression when in 1869, the first stirrings to revive the Lancashire unions took place. Initially, these moves received support from all quarters. MacDonald could not understand why other coalfields could have firmly established county organizations, with moderate leaderships, while Lancashire remained fragmented. He chaired the meeting to re-constitute the Lancashire and Cheshire miners at Manchester on 25 July, 1869. There was a good attendance of delegates: Wigan 54, Farnworth and Kearsley 32, Oldham 10, Worsley 10, Ashton-under-Lyne, 4, St Helens 2 and Poynton 1.[10] But from the start, deep local rifts appeared. The Bolton colliers regarded the Wigan colliers with suspicion, said James Lomax, because they worked longer hours, received poorer pay and produced more. To lead the discussion away from such a divisive topic, MacDonald spoke of the need for a united campaign for better mining legislation througout the whole country. However, this did not have the effect on conference he anticipated. For delegates seized on the idea of a national campaign, not limited to legislative questions, and considered the necessity of building a new national organization. MacDonald must have been horrified when James Lomax moved, and conference agreed, to dispatch organizers to Wales, Staffordshire and the North-East.

As Professor G. D. H. Cole put it: 'In 1868 the National Union called a conference of delegates from the various district Miners' Unions in Lancashire, with a view to creating an effective county organization. Instead the delegates formed the Amalgamated Association of Miners, which was meant to become a centralized national body organizing mine-workers over the whole country.'[11]

Paradoxically, national organization came first; county organization came second. Only after the Manchester conference had agreed to form the A.A.M. did its attention turn to the problems of Lancashire and Cheshire. Yet, at that time, the union's membership was extremely small—about 6,500, roughly a sixth of the total labour force of the two counties.

It may appear that it would have been more sensible to build a secure Lancashire and Cheshire base before venturing into other coalfields. But this overlooks the intense rivalry and localism displayed by district unions. They found it easier to unite when a firm national authority had been established, creating the conditions for local interests to be subordinated to the general will. Moreover, Lancashire, with its highly developed system of communications, was especially vulnerable to events elsewhere. A strike could easily be buried by a mass import of Yorkshire coal or by strikebreakers from Staffordshire. So the delegates at the Manchester conference were right to conclude that their welfare was inextricably linked to that of colliers elsewhere.

A month later, a second Manchester conference was held to hear reports of progress. In Lancashire itself, there had been a mixed response. A revival was taking place at Oldham, Ashton-under-Lyne and Duckinfield; Farnworth and Kearsley registered slight progress although many colliers still 'groaned in their slavery and desired deliverance but dare not join'; at St Helens however, a door-to-door canvass yielded only a few recruits.[12] Wigan, the district with the greatest membership, also displayed the greatest tendency towards local separatism. While it knew about the second Manchester conference—the date had been arranged at the first one—it decided to hold a district meeting on the same day. Apparently, Wigan thought its local business more important than the conference to consider the draft constitution of the A.A.M., which might be delayed a fortnight to suit its convenience. In spite of Wigan colliers' belief that 'other districts would do as they did', the Manchester conference continued without them.

Although delegates still came solely from Lancashire and Cheshire, the second Manchester conference did hear encouraging news from elsewhere. George Pickard had journeyed down to the Midlands. The colliers of South Staffordshire and Worcestershire, meeting at Dudley, had agreed to join the A.A.M., and appealed to their brothers in the Potteries to do likewise. Thomas Halliday persuaded delegates from fifty-two South Wales pits, representing more than seven thousand men, also to affiliate. But Robert Lewis, a Welsh-speaking miner who lived in Wigan, gave a dismal account of his expedition. 'There was not a single spark of union in North Wales,' he reported. There, they worked ten to twelve hours for 2s. 8d. daily. Even so, Lewis had been able to hold meetings of seven to nine hundred at Mold and other places. As a result, he had the names of men who wished to keep in contact with the new union.

The Manchester conference continued by considering the draft constitution for the A.A.M. It agreed that subscription to national funds should be threepence a month, with the executive committee empowered to levy all members equally in times of need. Rule 7 stipulated that all office-holders must be practising miners and fully paid-up members. The final authority and decision-making body was to be the national conference, held at least once a year but more often if necessary. In the intervening periods, a nine-man executive was responsible for day-to-day running of the organization.

The distinctive constitutional feature was contained in Rule 13:

> 'That all Districts support their own minor cases; but in case a general grievance arise throughout a pit, colliery or district, which cannot be settled amongst themselves, and is likely to lead to a strike or lock out, the president shall be warned of the same with all particulars, also the number of men likely to be affected thereby, when he shall give instructions for the executive to be called, who shall hear such grievances, and if the men be justified and receive the consent to resist such grievance then the executive shall have the power to award the support of the Association.'

Able to tap the financial resources of the entire membership in any serious industrial conflict, the executive could raise sufficient money to pay every man ten shillings a week, plus a shilling for every child under twelve. No provision was made for payment of wives, perhaps because the men who drew up the document were all Lancashire miners, familiar with the custom within the county of wives working in industry.

These provisions gave men greater confidence and security than hitherto. Almost as a natural transition, the second Manchester conference moved on from considering the constitution to discussing whether it should attempt to secure a wage increase. The moderates, led by William Pickard, thought the organization's membership and financial resources were still too weak; it was foolish to talk about issuing an ultimatum to the coalowners. Halliday, however, retorted, 'He found Mr. Pickard was always very careful, and coolly approached a subject, but this subject required not coolness but zeal.'[13] He did not see why Lancashire miners should be content with half-a-crown to three shillings a day when other workers were receiving five shillings. Conference agreed with Halliday and decided to ask employers to restore wages to the level prevailing before the last reduction.

The call for a wage increase aroused enthusiasm. Crowded meetings were held, and many colliers decided to join the Association. What seems originally intended to have been solely a Lancashire move became extended to other coalfields. The demand was officially lodged with the employers of Lancashire, Staffordshire and North Wales in November. In most districts, the owners, finding the market slightly more buoyant, granted a rise rather than have a strike. The Lancashire and Cheshire Coal Association conceded a 10 per cent increase on 7 December. In South Staffordshire, where colliers' wages were geared to those of ironworkers, the owners offered threepence a day. When 150 delegates, meeting at West Bromwich, rejected this offer and showed their determination to fight, the coalowners capitulated. Similarly, in North Wales, where Lewis, addressing meetings in chapels and schoolrooms, reported that he was now having 'complete success'. Not only had the last wage cut been withdrawn, but the masters had discontinued the practice of 'rucking', taking an extra hundredweight per 10 cwts. to allow for weight depreciation in the pile.

The A.A.M. congratulated itself on the success of its first venture. Speaking at Bolton, Halliday told a public meeting that 30,000 of the 40,000 Lancashire colliers had received an advance of half-a-crown a week (the equivalent of £221,000 per annum), 12,000 in North Wales had received a shilling a week and 30,000 in Staffordshire had received half-a-crown a week. Where owners resisted, Halliday made it plain that the union's entire strength would be deployed:

> 'There were 30,000 members in the Amalgamated Association, and if the large employers of labour—such as Messrs. A. Knowles & Son, Messrs. Pilkington & Evans, Messrs. Lee & Nightingale, and the Bridgwater Trustees—refuse to give an advance, he would recommend the men in their employ to strike, for a levy of one shilling per week upon each member of the Amalgamated Associa-

tion would bring in £1,500 per week or £15,000. in ten weeks which might easily be raised.'[14]

Halliday's threat led some firms to concede. Against those that still refused to pay the increase, the union took action, calling on their members to strike. But some colliers did not harken to the call. The Bridgwater Trustees granted a partial increase, 2½d. a ton and 3d. a yard instead of 3d. a ton and 4d. a yard demanded by the union. By making this concession, the Trustees forestalled the union. Only three hundred men struck; the rest remained at work; the strike quickly petered out.[15] Andrew Knowles adopted the same tactic, giving its colliers two-thirds of the union's demand, with equal success: its men, inured to servility, did not even protest. These two setbacks darkened an otherwise bright sky for the Amalgamated Association in Lancashire. These two big firms could still exercise a depressive influence on wages, not merely keeping their own below the prevailing level, but also acting as an incitement to the rest of the coalowners in their vicinity.

The A.A.M. had other problems. In North Staffordshire difficulties arose because of 'the mixed population settled in the districts. There were men belonging to Staffordshire, Wales and Ireland, and amongst these there was an entire lack of the confidence necessary in the establishment of a union.'[16] Besides their racial problems, the colliers of the Potteries were wary about doing anything that might offend the coalowners. Their recollections of the grim fate of previous unions and those actively associated with them, all aggravated the A.A.M.'s rather awkward position.

But Wales posed the biggest challenge. The type of response of many Welsh miners, encouraged by colliery officials, is exemplified by a letter that appeared in the *Wrexham Advertiser*. It claimed that crafty A.A.M. agents were coming into Wales intent on fomenting strikes and increasing costs 'to benefit the Lancashire collier and the Lancashire coalowners at the expense of the Welsh collier and his employer'.[17] Three-quarters of the coal used in Chester came from Lancashire, where coal was cheaper. The London and North Western Railway, which previously took Denbighshire coal, now secured its supplies from Lancashire. The letter suggested the sinister aim of Lancashire was to acquire the whole market of North Wales. Such arguments, playing on the Welshman's fear of the Englishman, were very potent. Doubtless, many could recall the riot at Mold in 1863 caused by the introduction of Lancashire miners who, it was claimed, were better at hewing cannel coal. 'The poor families who had to turn out of their homes were Welshmen. And they had to quit their homes to make room for Englishmen,' explained the *Wrexham Advertiser*.[18] This time, it appeared to many Welshmen, a slightly different strategy was being used towards the same end.

With slight variations, the South Wales coalowners trotted out the same arguments. They said they did not mind their workmen forming a national union of their own; it was joining an English-dominated body that they resented.

Despite such obstacles, by the beginning of 1870, the A.A.M. appears to have laid the foundations of a national union. In Lancashire, Staffordshire, Worcestershire, North and South Wales, it had created viable, if not thriving, organizations. Inevitably, its success raised the question of what should be the A.A.M.'s attitude to miners in better organized areas, like Yorkshire and the North-East, where the county unions were affiliated to the M.N.A.

Originally, it was not envisaged that the two bodies should be rivals. In August 1869, the M.N.A.'s Leeds conference praised the Lancashire men for forming the A.A.M., saying it would do a lot of good.[19] But when the M.N.A. leaders saw that the A.A.M. was not merely going to be a Lancashire association, like other county unions affiliated to it, a change of outlook and a growing hostility developed. Once the new organization adopted its strike strategy, with its centralized and more militant leadership, then the incompatibility of the two unions became obvious.

Nevertheless, the A.A.M.'s Wigan conference (January 1870) tried to reach an agreement. It sent a deputation to the National Association's conference, which was meeting in Manchester the following week. Opening the discussion was William Pickard who, as vice-president of the A.A.M. and treasurer of the M.N.A., had strong personal reasons for wanting to avoid conflict. He said:

> 'the recent conference in Wigan decided that a deputation should wait upon the National Conference to see what could be done with a view to general co-operation and the men of his district were exceedingly anxious there should not be two societies with one object. They were connected with both association and were very anxious, if possible, the two should work together or become one.'[20]

Then Halliday spoke:

> 'The financial position and the number of members of the

Amalgamated Association of Miners were very favourable and the objects they sought were the same as the National Miners' Association. The Amalgamated Society had not been got up in rivalry to the National Society; its formation was made imperative by the state of trade in Lancashire during recent times.'[21]

Halliday's conciliatory speech did not influence opponents. Implacable opposition came from Northumberland and Yorkshire, with Burt, Nixon and Normansell as the spokesmen. In essence, they objected to unity on three grounds. Firstly, because the organization in many coalfields was very weak. Evidentally, they had no desire to squander their resources on helping to get trade unionism established in places like Wales. Since they were not plagued with Welsh coal imports and blacklegs to the same extent as Lancashire, the effect would not be worth the reward. Second, they did not believe a national union should organize strikes. Besides conflicting with their strong tendencies towards industrial pacifism, they opposed nationally-led strike movements because of the ill-starred consequences of previous attempts. To venture on to the uncharted seas of class conflict might endanger their own prosperous and well-organized county organizations. There was also the danger—if Halliday had his way—of the creation of a highly centralized national body to which they would have to surrender some of their power and authority. The third objection, strongly voiced by Burt, was that the M.N.A. should strictly confine its activities to securing improvements in mining legislation. Any signs of militancy would jeopardize this objective. This point was reinforced by MacDonald, closing the discussion, who seized the opportunity to outline the progress in safety and ventilation regulations for which, he claimed, the M.N.A. was responsible.

There is no evidence that MacDonald put the issue to a vote. In a sense, the strong unions of Yorkshire and the North-East had settled the matter. They had made it plain that, whatever happened, they would not associate with the A.A.M. So, if the decision had gone against them, it would be tantamount to splitting the M.N.A. That is probably why MacDonald shrewdly did not put it to the vote. For only fourteen delegates attended the Manchester conference. Of these, eight came from unions that had already joined the A.A.M. (Wigan 4, Brierley Hill 2, Worsley 2) while the other six were from Northumberland (2), South Yorkshire (2) and West Yorkshire (2).[22]

In spite of the sparse attendance at Manchester, an indication that the M.N.A.'s fortunes had reached their nadir, it was nevertheless the fateful conference that took the decision—or, rather, took the decision by default. As MacDonald put it, 'After that they took to the left and we held our way to the right.'[23]

The A.A.M.'s rebuff led Halliday to develop a more thorough-going critique of the M.N.A. He told the Amalgamated Association's Wrexham conference (April 1870) that the delegation had been badly treated at Manchester. He conceded that the M.N.A. may have gained a little compensation for injured colliers, but it had squandered large sums on sending delegations to London. Little had been achieved. For instance, in recent negotiations MacDonald, Pickard and Normansell had abandoned the idea of demanding that practical miners should be appointed as sub-inspectors simply because the government opposed it. The M.N.A. was too half-hearted and conciliatory. The Wrexham conference endorsed Halliday's opinions and agreed that no further overtures be made.[24]

Many of these criticisms of the M.N.A. echoed those made five years before by the Practical Miners. The coalfields where support for the A.A.M. came from were the same as those that backed the Practical Miners. The industrial stances of the two organizations were remarkably similar.

A significant difference, however, between this and previous militant miners' unions was that W. P. Roberts does not seem to have been associated with it. His interest in trade unionism had not waned. He wrote a pamphlet on it and attended the 1870 Trades Union Congress, a few months before his death. The failure of W. P. Roberts to play a role may simply be a sign that, with the growing sophistication of trade unionism, there was no role for him to play. In the Miners' Association of the 1840s a freebooter like Roberts, a man with a forceful personality, could push himself to the fore. Callow, inexperienced colliers, lacking confidence in their own ability, allowed an outsider to become one of their leaders. But times had changed. 'The professional union official was, in fact,' as Professor H. A. Turner has remarked, 'by far the major innovation of mid-19th-century trade unionism.'[25] By the 1870s, the growth of union officialdom, encroaching on an ever-increasing amount of union activity had pushed Roberts, or any aspiring Roberts, out. Never again would an interloper play such a prominent part.

But union officials, *vis-à-vis* the rank-and-file, were not all-powerful. They could not do what they liked. The membership defined the limits within which officials could exercise their discretionary power.

William Pickard's predicament illustrates this point. By inclination and temperament, he had much more in common with the moderates of the M.N.A. than with the more militant Amalgamated Association. Yet he had no chance to persuade his district to disaffiliate from the A.A.M. and give the M.N.A. its undivided loyalty. Instead, he had to attend A.A.M. conferences, endure Halliday's criticism of himself and his only consolation was that he was able to get his district to remain affiliated to the National Association. But he had no illusions about where Wigan colliers' primary allegiance lay, a point they proved by disaffiliating from the M.N.A. in 1872.

William Pickard was able to persuade the Wigan district union to commit 'industrial bigamy' because the M.N.A.'s subscriptions were so small. The M.N.A.'s main function, that of securing legislative improvements, required minimal participation or finance. The colliers' role was primarily passive, to give discreet and respectful support to the deputations that went to London. Numbers were intended to impress legislators and it may be that, in order to have the most impact, the M.N.A.'s membership figures were deliberately inflated.

Evidence for suggesting this comes from analysing those districts which affiliated to both national bodies. At a M.N.A. conference in January 1871, it was claimed that the total membership was 110,000. Among these were Wigan 20,000, St Helens 10,000 and Ashton-under-Lyne 5,000.[26] These figures are difficult to reconcile with the returns published by the A.A.M. conference five months later. Then Pickard reported that Wigan's membership was only 3,000 while the St Helens delegate said that, although numbers had been as low as 137, they had now risen to 501.[27] No such comparison can be made for Ashton-under-Lyne as it never belonged to both associations. Even so, there is reason for doubting whether it had five thousand members since five years before, the Select Committee on Mines (1866) had been told that only three to four thousand colliers worked in the Ashton-under-Lyne district.[28] Moreover, at a union anniversay celebration at Ashton Town Hall in November 1872, Joe Booth, the local agent, claimed a membership of two thousand.[29]

Now, if the Wigan, St Helens and Ashton-under-Lyne figures are suspect, it raises doubts about the accuracy of the rest. William Pickard, of Wigan, was then the treasurer of the M.N.A. If he did not inquire too closely into the returns from Lancashire districts, he may not have looked very carefully into those from other areas. Another reason for questioning whether the M.N.A.'s returns were given with a high degree of accuracy is that they were all in round figures:

South Yorkshire	12,000	West Yorkshire	6,000
Northumberland	14,000	Wigan	20,000
Midloathian	7,000	Bristol	3,000
Ashton (Lancs)	5,000	Larkhall	9,000
Durham	18,000	St Helens	10,000
Fife	8,000		
	Total 112,000		

With the three Lancashire districts, it would appear that the numbers given roughly approximated to the total number of colliers working in the respective coalfields. Perhaps, the M.N.A. leaders thought that every miner would naturally want improved ventilation, safety legislation, etc., and hence they were perfectly entitled to claim to be speaking on behalf of the entire mining population. But it must be stressed that this method of calculating membership would not be suitable for the A.A.M. where up-to-date returns were vital for the calculation of levies and dues.

Comparing the two national miners' unions, Professor Cole cites the membership figures for 1873, when the M.N.A. had 123,000 to the A.A.M.'s 99,000. On the basis of these membership returns, not realizing that the concept of membership was different in the two organizations, he reaches the conclusion that the M.N.A. remained the larger body.[30] In any meaningful sense, it seems doubtful whether this was true.

The M.N.A. in fact, was a miscellaneous collection of mining unions, with markedly dissimilar power and wealth. Strangely, some of them were too rich to join the Amalgamated Association while others were too poor. In September 1873, a report reveals that the M.N.A. had fifteen affiliated bodies.[31] Of these, the combined balances of eleven of them totalled £34,000, exactly the same amount as South Yorkshire possessed on its own, while the richest union in the M.N.A., Durham, had £37,000. In contrast to these rich county organizations, the A.A.M. was relatively poor. Its entire balance at about the same time (A.A.M. half-yearly statement, October 1873) was only £20,324.[32]

It can be appreciated that rich county unions, like Durham, were reluctant to join the A.A.M. They feared that their impressive wealth, garnered through years of industrial harmony, might be squandered in an attempt to organize some distant coalfield. Moreover, the A.A.M.'s more militant attitude might disrupt the cordial relations with the coalowners, creating a conflict with perilous consequences.

The promptings of self-interest, self-preservation and organizational conservatism made the rich county unions shun the A.A.M.

The poor ones, on the other hand, had a different reason for remaining aloof. They were disinclined—indeed, they might be unable—to pay the large levies the A.A.M. called on its affiliated organizations to make. These could amount to as many as sixteen in a half year and were, in times of trouble, the main source of revenue. It was much easier and less exacting to belong to the M.N.A. where there were no levies and membership fees were small. The point is vividly made by this comparison: the M.N.A.'s total income in 1870, was £184 3s. 4d. whereas the Wigan union paid a levy of £288 a month during the Welsh strike of 1871.[33]

Notwithstanding this, in some areas there was no choice—they just had to belong to the A.A.M. The M.N.A.'s loose federal structure would not do for Wales, or even Staffordshire. The failure of previous attempts to establish colliers' union there proved it. If they were to rely solely on their own strength, defeat was inevitable. Only with sustained outside support could a union survive the unbending hostility of the employers. Consequently, in such coalfields a tight, centralized structure was a pre-requisite for existence.

So, when Lancashire, grappling with its own problems, formed the Amalgamated Association of Miners it struck a responsive cord elsewhere. Other colliers realized it fitted their requirements. Quite naturally, as a result of the differing circumstances encountered, new tactics were evolved, and these found expression in a national organization that rivalled the long established one of Burt and MacDonald.

CHAPTER SIX

The Success of Centralized Wage-Bargaining

In the early 1870s the prosperity of the coal industry created conditions propitious for the growth of the Amalgamated Association. The large profit margins, full order books, absence of coal on pitbanks helped to make the owners more amenable to the men's demand. They preferred compromises to clashes, even when these compromises had to be bought at a price. The shortage of colliers inevitably led to wage increases. But Halliday was not content to rely upon the blind working of market forces to improve his members' standards of living; centralized control of the A.A.M. would give him greater power to wring wage concessions. By deploying the full union's strength at strategic points, smashing employers' resistance in one area, he intended to make it impossible for coalowners to contest all-round increases elsewhere.

The history of the Amalgamated Association can be viewed as an attempt, extending over six years, to apply the theory of centralized control of union wage bargaining and the problems encountered in translating theory into practice.

In the initial phase, difficulties arose from organizational weakness. Recruitment was slow. After six months, the A.A.M.'s national membership had only risen to 12,000. Contact had been established with miners in Staffordshire and Wales, but these links remained tenuous. Men in these places had doubts of their own ability to sustain their unions when confronted with the coalowners hostility or of the A.A.M. nationally to provide adequate assistance. These misgivings could not be allayed by verbal assurances.

Wales, where wages were 20 to 30 per cent below the average in England, represented the most crucial challenge to the A.A.M. Unless a drastic improvement was achieved there, its entire organization was threatened: equalization of wages was essential for acquiring cohesion and strength. For if Welsh colliers remained poorly paid, they would be unable, or unwilling, to pay the heavy strike levies, which were a necessary part of the A.A.M.'s scheme. Usually, men on small wages are reluctant to provide financial assistance to workers in the same industry when their strike, if successful, will increase the differential. Moreover, so long as the disparity between English and Welsh wage rates continued, there would be a temptation for Welshmen to indulge in their traditional bad habit—crossing the border to take Englishmen's jobs during industrial disputes.

The Amalgamated Association resolved to explore the prospects in North Wales with a detailed investigation. When the fourth national conference was held at Wrexham (17-21 April, 1870), meetings were held in a large number of mining villages throughout the region. The result was disheartening, and confirmed reports given to conference. Lack of unity within districts, bitter rivalry between districts, an absence of militancy among men inured to poverty and servility, all represented serious obstacles to progress. The Rhos district, the biggest in North Wales, claiming 1,700 to 2,000 members, stated that it was too frightened of the coalowners to send delegates to the Wrexham conference. What was more depressing, colliers of Rhos, earning less than two shillings a day, were too poor to pay anything more than a nominal subscription. They were, consequently, granted a special dispensation. But, once this became known, the men of Cefn 'were annoyed because those in the Rhos district did not pay as much'.[1] The wrath of Cefn colliers had already been aroused. Whereas Lancashire had gained 10 to 15 per cent increases in the A.A.M.'s initial wages agitation, they secured only 5 per cent. Even this meagre amount was placed in jeopardy when miners from nearby Chick, hearing about it, assisted their masters to undersell the Cefn collieries. No wonder the Wrexham conference report stated that union organizer, 'Mr. Rone gave a rather dolorous account of affairs in North Wales.'

As prospects appeared so unpromising in North Wales, the A.A.M. decided to investigate the possibility of making its first major offensive in South Wales. There a small federal union, linking scattered groups in many of the valleys, had so far remained independent of the two national associations. But, fortunately for the A.A.M. the South Wales colliers quarrelled with Alexander MacDonald, thus giving it a greater chance of gaining converts. The dispute arose because the Welshman cherished the view that working miners should be appointed as sub-inspectors of mines. In 1868, they had gone to the length of writing a pamphlet, outlining the advantages of such an arrangement, and had even petitioned Parliament.[2] Naturally, the Welshmen strongly advocated sub-inspectors when the preliminary discussions took place before the drafting of the Mines Regulation Act. But MacDonald, Normansell and Pickard dropped the idea that colliers should be appointed as sub-inspectors immediately the Government made clear its own opposition. Both the Welsh colliers and the A.A.M. attacked MacDonald for his pusillanimity.[3]

The Amalgamated Association resolved to make South Wales a priority area, devoting as much union resources as possible to extending its influence. 'Agitation was rife,' reported a spokesman for the employers, 'professional unionists came into the district and roused workmen to a state of insubordination.'[4] Managers found it difficult to preserve 'the necessary discipline for working the mines to advantage'.[5]

Obviously, disputes were unavoidable, and there appears to be little evidence that the Amalgamated Association wished to avoid them. For each dispute was an opportunity to express, in practical terms, the support of the English colliers for the Welsh. The receipt of ten shillings a week strike-pay enhanced the A.A.M.'s popularity. As an Aberdare miner reported, these payments 'did more for the Amalgamation in this district than a hundred speeches; action speaks louder than words'.[6] Within a few weeks, three thousand new members had been enlisted in and around Aberdare. What happened there also happened elsewhere. It was not merely a question of financial assistance: in most of the struggles the A.A.M. emerged triumphant, and there could be no better form of publicity than success. Hitherto South Wales miners had, time and time again, experienced defeat. It appeared as if the Amalgamated Association had inaugurated a new era for them.

But the coalowners were not prepared to see themselves mauled, one at a time, by the A.A.M. One of the three big groups of coalowners—the Steam Coal Collieries Association—decided to make a united challenge. The time it chose could hardly have been more disadvantageous to the A.A.M. The union had spent £1,757 on disputes in South Wales, and there was only £770 19s. 0½d. in its coffers.[7] What was more, the A.A.M.'s membership at the Association's collieries was a mere 724, one in thirteen of the total labour force.[8] By inflicting a defeat on the union, the owners hoped not simply to wipe out the A.A.M.'s membership in its collieries but also to spread demoralization within its ranks throughout South Wales.

Halliday realized the magnitude of what was at stake. The steam coal owners had replied to the union's wage demand for a 10 per cent increase by announcing their intention of reducing wages by 5 per cent. If the owners were allowed to succeed, it would be a grievous blow to the union. Yet, for the blow to be warded off the A.A.M. would have to apply its central strike tactic, mobilizing all its strength behind the Welshmen. To do this, however, appeared extremely difficult. There was a difference between accepting the central strike tactic in theory and applying it in practice. Three pre-conditions were required for its successful operation: (i) that members would pay whatever special levies were imposed; (ii) that members would take an overall view, considering what was in the best interests of the whole, not their own narrow, sectional interests; and (iii) that those called upon to strike, obey the union's demand. In 1871, the existence of all three of these pre-conditions were in doubt.

The development of South Wales had placed a big burden on the small Lancashire membership. The frequent calls for special levies were unpopular. As early as April 1870, the Oldham district was giving as the reason for its failure to make new members the fact that men resented paying levies. Later, however, it even became the cause of an alarming exodus, colliers leaving the organization in protest. By 1871, St Helens' membership was 137, the lowest it had been since it began in 1863.[9] Farnworth and Kearsley were down to 1,800, about half the figure the district had at its peak. All told, of the forty thousand colliers employed in Lancashire only seven thousand were members of the A.A.M. in April 1871.[10]

At the same time, there were other disquieting signs in Lancashire. An attempt to gain a 10 per cent wage increase ended in failure. Some owners had granted the award in January 1871. But, by April, all of them, except those at Rochdale, had withdrawn the increase as it had become apparent that the union was not going to fight to bring the big companies into line.[11] All the union's available energy needed to be kept for struggles in Wales.

Failure to make headway made the Lancashire miners tetchy. Squabbles occurred within the ranks. The Manchester conference (May 1871), heard that Tyldesley had left the Wigan district and that Hollinwood, which had split off from Oldham, had lost 90 of its 170 members.[12] The proliferation of small district unions, unable to discharge their functions adequately, could not fail to affect adversely the whole Amalgamated Association.

It was symptomatic of the tendency to place narrow, sectional interests first. District unions with pressing grievances did not see why they should relinquish the right to obtain redress merely because it failed to coincide with the overall strike strategy. Halliday's own district was among the foremost critics. Wages in Farnworth and Kearsley were below the Lancashire average, and colliers believed they were entitled to withdraw their labour to achieve parity. When they heard that no strike pay would be forthcoming if they did, 'the executive were brow-beaten by the men; and a different plan was followed to that recommended'.[13] Both sides felt aggrieved, and so the matter was referred to the Wrexham conference where, after a heated debate in which Halliday said Farnworth and Kearsley had broken the constitution, it was agreed to leave the question of strike strategy to the executive. The resolution was only narrowly carried, 17 votes to 15. Significantly, at successive conferences after Wrexham, the rank-and file challenged the executive's prerogative to determine strike policy, but it never won the vote. Even so, the excutive was far from being in complete command: colliers sometimes relieved their frustration by resorting to unofficial action.

Paradoxically, while the Amalgamated Association leaders had immense difficulty in restraining some of their members from striking, they had equal difficulty endeavouring to induce members to strike when it wanted them to do so. When the South Wales colliers saw that they were on a course that would lead to a head-on collision with the mighty Steam Coal Collieries Association, many of them had a failure of nerve and wished to withdraw. All previous contests had ended in dire disaster. A six-week strike in 1856 had finished in defeat, with the union smashed and its leaders forced to leave the country. Then, again, in May 1864 a nine-week strike at Abercwmboy had failed and there had been general demoralization. 'No isolated strikes have occurred since', reported the secretary of the employers' organization.[14]

Thomas Halliday journeyed down to Mountain Ash, on 30 May, 1871, for a conference of Welsh colliers. He wanted to imbue the men with a spirit of confidence, a willingness to fight. He promised them that, if they went on strike, thirty to forty thousand men would pay levies to support them. As the A.A.M. at the time only had 23,901 members, there appears to have been some substance in the allegation of the *Western Mail* that he engineered the strike.[15] The promises of financial support made by Halliday at the Mountain Ash meeting were beyond the A.A.M.'s resources to fulfil and they were not fulfilled.[16] In spite of his personal reputation and assurances, the meeting appears to have remained very divided: of the 18,326 colliers represented at Mountain Ash only 4,940 wanted the union to stand by its original claim for a 10 per cent increase; 7,147 were prepared to settle for a five per cent compromise.[17]

The strike, which began on 1 June, 1871, took place in the context of nationwide rising coal prices. Hence a strong temptation existed for owners to arrive at individual agreements with the union. Indeed, seven masters, employing 2,351 men, did that right from the start. After a fortnight, a further two coalowners, employing 1,427 men, also made their separate peace.[18] Besides such defections within their own ranks, the coalmasters were bedevilled by the activities of other employers. The Steam Coal Collieries Association's membership only produced 75 per cent of the total output of that type of coal.[19] By stepping up production, non-affiliated owners could make large profits. Similarly, with the ironmasters: before the strike, they had promised not to engage any men in their mines who had worked for members of the Steam Coal Collieries Association. But, with rapidly rising prices, the desire to break their promise proved irresistible.

The influx of strikers into other collieries lightened the burden on the Amalgamated Association. From having 9,222 claimants for strike pay, the number was down to 3,882 by 29 July. At the same time, employers involved in the dispute must have found it galling to see others profiting at their expense and even have experiences like this: 'The owners of a neighbouring colliery sent to the Manager of the Ocean pits, asking for the loan of horses, in order to keep at full work the new hands lately engaged by them, many being the strike hands from the Ocean pits!'[20]

The Steam Coal Collieries Association had hoped to minimize loss of output through the mass employment of blacklegs. It approached Captain W. Godden, an unemployed manager of a Cornish tin mine, offering him a pound for each recruit. But enthusiasm among the Cornish and Devon men quickly evaporated when they heard of the high casualty rate in South Wales. Paul Roper, a veteran at enlisting strikebreakers, sought to engage men from South Staffordshire. But, as a result of the co-ordination between districts of the A.A.M., his task was more difficult than even before. Wherever he went union officials were there to thwart his endeavours. Meetings were held throughout South Staffordshire to persuade men not to blackleg. In triumph, the officials of the South Staffordshire district could write to the men of the Rhondda and Aberdare valleys: 'Respectable men will not come into South Wales; nothing but the very scum of the country will put their feet upon your soil.'

The employers admitted that only a trickle of strikebreakers, the roughest and most unskilled of men, came to South Wales. Because of their poor quality, coalowners were reluctant to make permanent arrangements for them to work in the pits. Instead of evicting strikers from their cottages, owners accommodated the newcomers in schools, repair shops and even lavatories.[21]

When the Staffordshire men began to work—'or rather pretended to', as an owner's spokeman put it—production was wretched. To complicate matters, on 2 August, firemen, 'generally supposed to be most reliable officers of the pit, expressed their sympathy with the old hands in a practical manner, by handing in their lamps, and refusing to go down the pit with strangers.'[22] Likewise surface and underground overmen showed lukewarm endeavour to help their employers.

On 14 August, the experiment collapsed completely: 'A large party of Staffordshire men, after fraternizing with the old colliers, left the Lower Duffryn and Navigation Pits, and returned home. During the next few days, the remainder left.' The men complained that they had been deceived by Paul Roper, who had talked about the high wages they would receive in South Wales but had never mentioned the strike. They also objected to being heaped together about the collieries, guarded night and day by the police. A further reason, in some cases, may have been that employers tried to tighten discipline, restricting the parties and drinking, as the amount of work done had been 'miserable in the extreme'. But, whatever the causes, the departure of the Staffordshire men was a grievous blow to the owners. As their spokesman put it, 'A more disastrous event could not well have occurred at the juncture.'[23]

Prospects looked bleak for the Steam Coal Collieries Association. Not only was its output nil while its competitors profited from its misfortunes, but the Association was growing more unpopular. Frequently, the union spokesmen had expressed their willingness to negotiate whereas the owners had not. So blame for the suffering caused by the dispute was largely held to be the responsibility of the employers. The *Western Mail* wrote, 'the position assumed by the men is one which deserves and receives the sympathy of the general public'.[24]

The Strike, which ended after twelve weeks, on 24 August, resulted in a settlement that was substantially a victory for the A.A.M. The union had successfully resisted the demand for a wage cut; it had repulsed an attempt to smash its organization; it had gained a 2½ per cent wage increase. As Mr E. W. Evans observed, 'The terms of the award were less important for the union than the fact that Halliday had been admitted to the arbitrator's meeting. This constituted recognition of the organization by the coalowners, which was an obvious achievement.'[25]

The lessons of South Wales were not lost on employers elsewhere. The cost had been extremely high for the Steam Coal Collieries Association. Its total output in 1871 was 1,702,061 tons as against 2,229,530 tons in 1870, a drop of 527,469 tons. But their losses did not end there: the strike, by dispersing its skilled labour force, had long-term repercussions. Production only slowly crept back to the 1870 level, so the owners did not participate, to the full extent, in the ensuing prosperity of the industry. It took years for them to recover fully.

A consequence of the South Wales strike of 1871 was that the Amalgamated Association was treated with greater circumspection by coalmasters throughout the country. They were less prepared to adopt a bellicose attitude. Far better grant wage concessions than indulge in all-out class war which, as well as being very costly, might be lost. So, in the aftermath of the South Wales struggle, the A.A.M. was able to secure larger wage awards for its members than had previously been the case and, usually, this was achieved more easily. In part this was a recognition of the union's great acquisition of strength; it was also a result of the boom being experienced by the coal industry. What it represented was a turning-point in the fortunes of the Amalgamated Association.

In its militant quest for wage increases, the A.A.M. continually expressed its readiness to take, whatever the claim, to arbitration. This might, at first glance, appear strange: the champions of arbitration are customarily thought to be the moderates, not the militants. But such a tactic had distinct advantages for the Amalgamated Association. Sometimes it was a means of gaining union recognition, as it was in South Wales. Also, it helped the A.A.M. to acquire support among the general public. By expressing its willingness to settle differences round a table, the A.A.M. was able to pose as the epitome of sweet reason; the onus for any strike appeared to be placed on the shoulders of the employers. Moreover, by proposing that the dispute be sent to arbitration, the A.A.M. gained time for reflection: the union could calculate what strain the strike would impose on its financial resources; the employers could ponder the damage to their profits. But second thoughts do not always lead to the resolving of differences nor do professions of peace always coincide with peaceful conduct. An employers' representative, Fothergill, chairman of the South Wales Ironmasters' Association, viewed the A.A.M.'s tactic in a different way: 'There were men amongst the miners whose mouths overflowed with arbitration and conciliation, but whose hearts were full of bitterness and discords.' Dr V. L. Allen tends to confirm this opinion, 'In addition to a large number of local strikes, the Amalgamated Association of Miners was rarely left without a major strike or lock-out to finance.'[26]

The Liberal politician, A. J. Mundella, first mooted the idea that arbitration should be applied in the coal industry. During the 1868 strike in Lancashire, he wrote to the Mayor of Wigan suggesting an arbitration court should be created.[27] Initially, the owners refused to co-operate with the scheme, but eventually, after the strike had dragged on for a few more weeks, they agreed. The settlement reached gave the owners the 15 per cent cut in workers' wages that they had wanted all the time. It did, however, also stipulate there should be no victimization of trade unionists and, once 'the agitation subsided, the employers will seriously take into consideration the practicability of establishing a Board of Arbitration and Conciliation'.[58]

The Lancashire coalmasters had qualms about creating such a system and, after considerable deliberation, the idea was dropped. There was the feeling that it would be wrong to have interlopers, people unconnected with the coal industry, being the ultimate arbiters on such a vital matter as wages. It was generally thought that it would be better if market forces, the interaction of supply and demand, were left to resolve questions of prices and pay.[29] Mundella sought permission to address the Lancashire and Cheshire Coal Association, hoping to persuade the owners to change their minds. He told the Association he thought arbitration offered three advantages: first, resolutions were generally obeyed, thus giving both sides greater security; second, that trade union subscriptions tended to be smaller; and, third, that strikes, costly to masters and men alike, would be avoided.[30]

The *Colliery Guardian* expressed doubt about Mundella's views. It thought that arbitration would be unable to preserve peace within the industry and that, merely to strive to introduce it once a dispute had begun, would make arbitration a weapon in the hands of one side or the other.[31]

The activity of the Amalgamated Association tended to confirm the opinions of the *Colliery Guardian*. Definitely, Mundella's second and third reasons for advocating arbitration were not borne out in practice. The A.A.M. lived a life of struggle, probably having more strikes in a period of six years than any other mining union. Also, because of its aggressive stance, it made larger calls on its members' pockets than any previous union in the industry.

But there seems to have been some substance in Mundella's first point. After arbitration agreements had been reached, they were usually honoured. In the case of the South Wales steam coalminers, this was a distinct disadvantage. For the agreement, was settled before coal prices, which were gently rising, had begun to soar. Within a few months of their leaders' signatures being appended to the document, the men would have been able to secure a much higher wage:

> 'On a rising market the men's bargaining position would have been very strong. All that stood between the employers and a universal agitation for higher wages was the award. And only the influence of the union leaders prevented the colliers from repudiating it.'[32]

Halliday, who had been nicknamed 'the prophet', lost his popularity among the Welsh steam coalminers but, despite the criticism from the rank-and-file, he did not swerve from his view that the agreement should be honoured.

Mr E. W. Evans suggests that there were four reasons for the A.A.M. adopting this position: first, that since the agreement tied the wages of colliers to those of ironworkers, it made future pay changes automatic and beyond the coalowners' control; second, that the strike had gravely depleted union funds; third, that the union had been implicated in the decisions reached; and, fourth, that 'to strike would have alienated public sympathy, lowered the organization's prestige, and undermine the integrity of the executive'.[33]

Undoubtedly, there is considerable truth in Mr Evans' reasons, but in my opinion they are not the whole story. It was essential for the Amalgamated Association to fight only limited battles, strikes confined to a particular coalfield, while it preserved peace elsewhere. Absolutely necessary, as a means to achieving this end, was to be able to reach agreements with coalowners that would be respected by both sides. Had the A.A.M. broken the South Wales agreement immediately it turned out to be disadvantageous, it would have been an open incitement to coalowners to do likewise whenever an agreement began to conflict with their self-interest. Instead of having industrial peace in most of the coalfields, there would have been the constant threat of war, a general state of insecurity and conflict which would have made it impossible for the A.A.M. executive to have applied its centralized strike tactic in limited sectors.

Besides this general point, the A.A.M. leaders were probably pushed to adopt a position of caution because of the overall condition of the union. Miners in Lancashire and elsewhere were already restive about the large sums that had been levied to help the South Wales men. To have another strike there, with the concomitant financial burdens it would place on colliers in other coalfields, might have placed an intolerable strain on the cohesion of the union. There are limits to the amount of money men will outlay without seeing any direct, tangible results. In the circumstances, the A.A.M. leaders change of front appears quite understandable: Lancashire had been restraining its efforts for bettering itself so that all available resources could be devoted to South Wales; now it was the time for South Wales to show restraint so improvements could be gained in Lancashire.

Preliminary preparations for the union's push in Lancashire had been laid well in advance. A general meeting of miners from all over the county, representing about a hundred collieries, met at Hindley on 24 April, 1871 to discuss wages. This gathering, known as 'a pit-set conference', was open to unionist and non-unionist alike. It seems to have been a feature of the Lancashire coalfield, and this one at Hindley appears to have been the first to have been held. The advantage of pit-set conferences was to obtain a consensus of miners' opinions and, perhaps, upon the basis of this, to achieve greater coherence and unity of action. Instead of non-unionists being, as is generally the case, apathetic or even hostile to the union, the holding of pit-set conferences helped to create a *rapprochement*, which could be useful. For example, in the North Staffordshire strike of 1883, as Mr Page Arnot observes with some surprise, Lancashire union members levied themselves for the Staffordshire union members while the non-union members levied themselves for their non-union brothers there.[34]

But pit-set conferences had also their disadvantages. They were attempts to get large groups of men to behave as if they were union members without affording them full union protection. A man, victimized by his employer for implementing a pit-set conference decision, could not expect financial help from the union's coffers. There was, moreover, much greater uncertainty about the carrying out of conference resolution than is usually so with ordinary union conference. A decision of a pit-set conference merely had a moral authority; the union could apply no effective sanction against non-unionists who failed to implement the decision. But there was for the union another danger: non-unionists, not having to worry about the state of the union's funds, might get a pit-set conference to adopt a more militant line, one that jeopardized the union's very being. Indeed, this was one of the reasons—that, in the eyes of the union leaders, pit-set conferences were prone to recklessness and iresponsibility—that led to their abolition in the 1890s (see Chapter Fourteen). And, of course, by then the need for them had diminished: a greater proportion of colliers were union members.

But in the 1870s organization in Lancashire was much more primitive. The Hindley gathering was informed that only seven thousand of the forty thousand were members.[35] The lack of co-ordination and contact was vividly illustrated. The Hindley conference welcomed the attendance, for the first time, of delegates from North-East Lancashire. In welcoming them, Thomas Halliday said he neither knew their district nor had visited it. This, surely, clearly indicates the pre-occupation of the better organized parts of the county with menace from other coalfields, the threat that their struggles could be crushed under the weight of coal and strikebreakers imported into Lancashire. Consequently, they had been more concerned with what happened in Staffordshire and South Wales than in North-East Lancashire. Halliday visited South Wales many times; his first visit to Burnley was in 1873.

In a sense, therefore, the institution of pit-set conferences was an attempt to achieve an all-embracing, if less strong, unity among the men to compensate for the small membership, the fragmentation, the semi-autonomous district unions—features that characterized Lancashire but were not so prevalent elsewhere.

The Hindley pit-set conference resolved to begin a concerted wage agitation throughout the county. Significantly, soon after the capitulation of the South Wales Steam Coal Association in August 1871, the Lancashire masters conceded a 10 per cent rise. A second increase, again of the same amount, was secured in January 1872. This brought most Lancashire districts up to the previous highest wage peak, that secured in 1867. (See Table 4.)

Then, the floodgates were opened to unprecedented increases. They came in rapid succession until, by April 1873, wages had risen from between 50 to 70 per cent, depending upon the coalfields, above the 1869 figure. In this wage-rush Lancashire districts were to the fore. Whilst traditionally the county is a low-wage area, by April 1873 three Lancashire districts—Wigan, St Helens and Oldham—were the highest paid colliers in the Amalgamated Association. (See Table 5.) This probably cannot be attributed to the national union leadership, almost entirely Lancastrian in composition, deliberately devoting the major portion of resources to struggle in the county so much as the Lancashire coalowners having to acknowledge a changed relationship of forces. Stripped of their normal weapon, the mass importation of Staffordshire and Welsh blacklegs, they had no alternative but to make concessions.

Of course, colliers were overjoyed with their new-won prosperity. They showed their gratitude by flocking to join the Amalgamated Association. In the six months, April to October 1872, membership rose from 23,676 to 70,536. It appears to have reached its peak at the Manchester conference of April 1874, when the union had 106,000 members. In a biographical article on Halliday, *The Beehive* claimed that it had reached 180,000, but this is not confirmed in any of the A.A.M.'s conference reports.[35] Nevertheless, *The Beehive* was clearly right in realizing the key to success lay in its ability to vanquish obstreperous coalowners. Speaking of the second South Wales strike, against the ironmasters' collieries in 1873, *The Beehive* wrote:

> 'Clearly this great power was due to the first place to the power of amalgamation, which brought the undivided force of so vast a body of men to operate at one moment, at one point, for one object; and secondly, to the steady, judicious, skilful, and determined manner in which Mr. Halliday, as president, conducted the struggle.'[36]

Unlike during the first big strike in South Wales, when the A.A.M. failed to keep its promise about strike pay, in the 1873 dispute not only was strike pay paid in full but also assistance was given to the non-unionists as well.[37] The significance of this cannot have failed to impress the coalowners. They were confronted with an organization which, while it kept its subscriptions small, had an immense capacity to levy its members. The six-month balance sheet, presented to the A.A.M.'s half-yearly conference at Newport (Mon.) on 1 April, 1873: contributions were £5,855 whereas special levies brought in £43,831. Twelve levies had been made during that six-month period and, at the end of it, the union still had a balance of £22,861, compared to £4,524 at the previous conference. To have been involved in such a hard-fought struggle as that in South Wales and emerge stronger than it went in: the A.A.M.'s strength must have been a source of concern to the coalowners. As one of them put it to the Royal Commission:

> 'It is quite clear that the perfect system of organisation, which the men have got amongst them, unless it be met by some counter-organisation, must be eating up piecemeal the coalowners and capitalists.'[38]

While the coalowners expressed their anxiety, the colliers expressed their pleasure. Newspapers of the time are full of reports of colliers flaunting their wealth. For instance, the *Manchester Guardian* wrote: 'St Helens at present is a veritable Elderado for workmen, a Tom Tiddler's Ground wherein by working seven or eight hours a day they can pick up enough gold and silver to enable them to treat the coppers with contempt.'[39] It said men took greater care with their appearance: young men wore 'clean linen, smart neckties, gorgeous "belchers" round their throats, and brightly polished boots'. Half of them, the newspaper claimed, smoked cigars. The sole topics of conversation were dogs, pigeons, and wrestling. Other newspaper reports talked of the much high incidence of absenteeism and drunkenness. They even mentioned colliers drinking champagne, although Pickard claimed that only about five out of every hundred had ever tasted the drink. Nevertheless, however much newspapers and the envious might have exaggerated the colliers' affluence, there can be no doubt that it did exist.

Table 8 Average Collier's Wage in South Lancashire Coalfield

1867	125
1868	110
1869	100
1870	110
1871	120
1872	Jan., 125; June, 145; Sept., 160
1873	March, 170
1874	March, 155; Sept., 140
1875	130

Base year 1969 = 100

Sources: Select Committee on the Scarcity and Dearness of Coal, *Manchester Guardian*, 18 April, 1873, and other newspaper reports.

Table 8a Wages and Prices, April 1873

	Average daily pay				*Cost of coal at pitbank*		
1	Wigan			1	West Bromwich	25s.	6d.
	St Helens	6s.	9d.	2	Cannock	25s.	od.
3	Oldham	6s.	8d.	3	North Wales	23s.	6d.
4	Swansea Vale	6s.	6d.	4	Neath		
5	W. Cumberland	6s.	3d.		Oldham		
6	West Bromwich	6s.	1d.		New Tredegar	22s.	od.
7	Neath	6s.	od.	7	Swansea Vale	21s.	6d.
8	Cannock			8	North Staffordshire		
	Farnworth & Kearsley	5s.	10½d.		Aberdare	21s.	od.
10	North Staffordshire	5s.	3d.	10	West Cumberland		
11	Aberdare				St Helens	20s.	od.
	New Tredegar	5s.	3d.	12	Farnworth & Kearsley		
13	Worsley	5s.	od.		South Staffordshire	19s.	od.
14	North Wales	4s.	10½d.	14	Wigan	18s.	6d.
15	Abertillery			15	Radstock	18s.	3d.
	South Staffordshire	4s.	6d.	16	Abertillery	18s.	od.
17	Radstock	4s.	1½d.	17	Worsley	17s.	od.

	Average daily pay				*Cost of coal at pitbank*		
18	Bristol	4s.	1d.	18	Salop	15s.	10d.
19	Burnley	4s.	0d.	19	Bristol	15s.	0d.
20	Salop	3s.	9d.				

Figures taken from delegates' reports to Newport conference, 1 April, 1873.

Mean average given where delegates reported variations, e.g. Wigan wages 6s. 6d.—7s. 0d. equals 6s. 9d.

Burnley district did not give figures for cost of coal at pitbank.

CHAPTER SEVEN

The Downfall of the Amalgamated Association

Industrialists and domestic consumers grumbled about the rapid rise in the price of coal. In London the average price had risen from 18s. 6d. a ton in 1870 to 32s. 6d. a ton in 1873. Inevitably, this caused hardship in many homes and higher production costs in factories, thereby affecting the competitive position of British industry in world markets. The trouble was aggravated by the rapid rise in the use of the commodity: consumption per head had increased by 65 per cent between 1858 and 1872. Whereas 64,587,899 tons had been consumed in the former year, it had swollen to 123,386,225 tons in the latter. So, obviously, on the whole community the burden of higher prices was one that was difficult to bear.

Problems were compounded because there was no alternative to paying the higher prices, regardless of how high they rose. The scarcity of wood, the only other fuel possible for the domestic consumer, made it impractical for householders to make any significant switch from one to the other. Similarly, industrialists were left with little or no manoeuvrebility. Their use of coal flunctuated with the tempo of industrial activity. They had no alternative but to submit to higher coal prices if they wanted increased output. Unlike the present time, when there are other sources of power—electricity and oil, for example—in the eighteen-seventies dependence on steam necessarily involved coal. Consequently, it was impossible for there to be what economists term a substitution effect, decreasing the consumption of one commodity when its price rises and replacing it by consumption of another.

In an endeavour to avoid this harsh fact, desperate industrialists tried many ingenuous stratagems. The London and South Western Railway, for example, mixed chalk with the coal, all to no avail.[1] Really the only possible method of cutting coal consumption was to introduce more efficient methods of utilization. The tendency for this to happen had existed for some time. In the third volume of his book '*Capital*', Marx observed:

'The machine power of the district of Lancashire was for almost

THE MINERS' STRIKE AND THE REASON WHY.

A good cause and a trusty hand,
  A stout heart and true,
And our masters shall understand
  What mining men can do.
And will they have the fifteen per cent,
  Or must we idle lie,
Then our masters shall repent
  And we'll show the reason why.

CHORUS.

What will they scorn our rights, and then
  Must we of hunger die.
Then twenty thousand mining men
  Will know the reason why.

he miner he is strong and bold,
  A gallant toiler is he;
e toils in dark chambers in the earth's hold
  And do ye grudge his fee?
e toils hard where he can't stand,
  And danger is no stay;
at death and danger is at every hand,
  And do ye grudge his pay?

ust we in vain for justice call,
  Then let man to man be true,
d cry, Come forth ye tyrants all,
  We are better men than you.
e miner he is strong and bold,
  Then must he idle lie,
ll they his just wage withhold,
  And show no reason why?

They talk of the hazards of capital,
  And shun all the moil
Of a fearful trade, ye cowards all,
  Will you underpay its toil?
Then will you have fifteen per cent,
  Or must we idle lie,
If we do starve you shall repent,
  Or we'll show the reason why.

We offer ten, that's more than enough,
  If masters were wise and knew
That our hearts like our hands are tough
  And our course is sound and true.
The miner he is strong and bold,
  Then shall he idle lie,
Shall his strength to a greedy crew be sold,
  Must he of hunger die.

We give hard toil, we ask fair play,
  We hazard life and limb,
Amid a thousand deaths we toil away
  In the earth dark and grim.
Then must we lose fifteen per cent,
  Or must we idle lie,
Our tyrant masters may yet repent,
  Or we'll know the reason why.

Songs such as these were sung by miners during industrial disputes and after pit disasters. Besides helping to maintain morale, they also elicited sympathy—and sometimes money—from the general public.

COLLIERS' NEW HYMN.

Each feeling heart pray lend an car
Unto this mournful tale;
To draw a tear of sympathy,
I'm sure it cannot fail.
A number of these poor Colliers,
I grieve to have to say,
By fire-damp, near Barnsley,
Were lately swept away!

God of my life, now I descend
To labour under ground,
While danger all my steps attend,
And darkness all around!
The fiery damp thou can'st control,
And stop my fleeting breath;
And cause that heavy fall of coal
To crush me soon to death.

To face these horrid dangers,
Some brave men did descend,
Unto their fellow-creatures
They did assistance lend;
It struck their hearts with terror,
And filled them with dismay,
To see so many lifeless,
In death's cold arms that day.

Oh God! our lives are in Thine hand
To spare or take away;
And to Thy care we'll recommend
Ourselves throughout this day!
From sudden death deliver us,
And let thy will be done:
I trust, oh God! thou'lt let us see
Once more the setting sun.

This was a dreadful accident,
From all accounts we learnt;
Their parents scarce could know their own,
They were so badly burnt:
We hope the Lord received their souls,
In heaven above to dwell;
The sufferings they did undergo
No mortal man can tell.

Many go down to rise no more
To see the glorious light;
Perhaps, before the day is o'er,
They enter death's dark door:

Parents mourn, and children weep
For those who are laid below;
Fathers and sons in death's long sleep,
Crush'd by that fatal blow.

With hearts so light they left their homes
Upon that fatal morn,
They little thought of their hard fate,
Before that night did dawn;
Our lives are quite uncertain,
As plainly you may see,
One moment we're in health and bloom,
Next in Eternity.

O, give us grace, our Lord, to see,
That all there may be well!
Bless God, when we descend the pit,
'Tis not the pit of hell!
Thou can'st create, thou can'st destroy,
Whatever thou thinkest best,
If it's thy will this day we die,
Then take our souls to rest.

Their money, colliers dearly earn,
They should protected be,
You all well know a collier
From danger's never free;
The cries of wives and children
Were dreadful, as we find:
May God protect their lovely babes,
Whom they have left behind.

The cries and groans of those poor men
Would cause your hearts to ache;
And many shed sad tears that day
For those poor colliers' sake.
Think of those poor mournful widows,
Whose hearts did ache with pain,
To think that morn would prove the last
They e'er should meet again.

Now their wives and helpless children
In sorrow are left behind,
May the Lord be their protector,
And unto them prove kind!
For grief is so distressing,
So piercing to our ear,
Each joy is fled, their father's dead,
And those children left to rear.

forty years under the pressure of timid and prejudiced traditions. But now the engineers have been happily emancipated. During the last fifteen years, but particularly in the course of the last four (since 1848) a few important changes have taken place in the operation of condensed steam engines. The result was that the same machines accomplished far more work, and the consumption of coal was considerably decreased at the same time.'[2]

Marx mentions the pioneer work of the famous engineer James Nasmyth, of Patricroft, and cites an instance where a machine had a performance of one horse-power per hour for $3\frac{1}{2}$ to 4 lbs. of coal instead of the previous machine that used 12-14 lbs. of coal.[3]

The trend that Marx discerned in the eighteen-forties became much more pronounced during the eighteen-seventies. The Select Committee on the Scarcity and Dearness of Coal heard many witnesses mention the introduction of more efficient boilers. Alfred Hewlett, of Wigan Coal and Iron Company, reported that this was becoming quite a common practice in Lancashire.[4] The Mines Inspector, J. Dickinson, said that textile manufacturers were, as a result, achieving economies of about 30 per cent.[5] With such an inducement, almost every employer wanted new boilers. But the limited capacity of that section of the engineering industry meant there was a bottleneck, with many unsatisfied customers. In the short-run, for the majority there was no way out, no boilers available, and hence they had to pay the higher coal prices.

But the coalowners themselves were in a very similar predicament. With higher labour costs, they had a great incentive to substitute capital for colliers. But the sudden increase in demand for colliery machinery created for them, too, a bottleneck. Even where machines were readily available, however, coalowners had to be cautious about their introduction. Many of the coal-cutters were fragile, and liable to breakdown. Dickinson reported that, in his journeys around Lancashire collieries, he often asked owners with coal-cutting machines how they were getting on, only to receive the answer, 'Oh, it is under a fall of roof' or 'It is not working today; the roof has fallen on the air-pipe.'[6]

As William Frith pointed out to the British Association, coal-cutting machines were peculiarly prone to falls of roof.[7] Since they were dependent for their source of power upon steam, the coal-cutters tended to soften the roof immediately above them. There was, of course, the added danger that the introduction of steam engines would increase the likelihood of explosions.

In spite of these hazards, coalowners were eager to instal coal-cutters. Frith clearly reveals the reason. In his speech he said that his machine normally cut 60 yards a shift, 'being equal to a day's work of twelve average men'. But this calculation was subsequently challenged by James S. Jeans, who claimed that Frith's machine resulted in a saving of only 8·42 per cent.[8] Inventors frequently exaggerated the performance of their machines. Of the 104 coal-cutters patented, Jeans said less than half-a-dozen had proved successful.

So, in the short-run, they proved to be very little help to coalowners. Too frequently, great expectations were followed by great disappointments. Even so, with the passage of time, they became more reliable. By 1875 the *Colliery Guardian* was cautiously recommending them. Its reasons appeared to be similar to those advanced by W. J. Clapp, in his pamphlet 'Improved Machinery and Apparatus': the disturbed state of the labour market and the improbability of bringing wages down to the pre-1869 level made the spending of £150 to £200 on a coal-cutter a worthwhile investment.[9]

In the long-run, the advantages of coal-cutters began to outweigh the disadvantages more decisively. They had, as H. S. Jevons pointed out, six merits: they produced (1) a larger percentage of big lumps, in firmer and better condition; and (2) a more regular line of coal-face, which facilitated better timbering and ventilation. This afforded conditions for (3) more regular and systematic working, as well as increasing safety. (4) Seams which, because of thinness or hardness were previously unprofitable, could now be worked. (5) Holing had to be done less frequently. And, the overall effect was (6) to increase output and improve its regularity.[10]

Coalowners utilized other mechanical aids. The most helpful, as a means of economizing with labour, were improved haulage arrangements. Dickinson described to the Select Committee what happened:

> 'In Lancashire, where we have so many steep mines, machinery is applied in almost every instance where it is possible, and there is hardly a ton of coal that comes to the surface, in a large number of mines that has not been moved once or twice, sometimes three times, by machinery.'[10]

The growing use of machinery made coal mining a more capital-intensive industry. In 1874, machines on average cost £40 per horse power. They became a necessity not merely for haulage but also for winding. Lancashire had some of the deepest workings in the country. Rose Bridge, near Wigan, reached 816 yards in 1869 while a shaft at

Ashton, Manchester, was 896 yards deep by 1881. Workings at such a depth required cages, with loads of up to four tons, travelling at nearly a mile a minute. This, as Boyd pointed out, necessitated the use of 1,500 horse power winding engines, extremely expensive pieces of equipment.[11]

The sinking of the shafts was also highly costly. In 1874, the *Wigan Observer* stated the sinking of a single shaft cost between £40,000 and £100,000.[12] Eight to ten pounds a yard would have to be spent on preliminary borings.

Need to employ more capital accentuated the growth of larger production units. Wigan Coal and Iron Company's output reached two million tons in 1873. Pearson and Knowles, with an average output of 800,000 tons per annum, was estimated in the same year to possess a million pounds of capital.[13] Companies such as these were the giants of the industry. They exemplified the trend towards large scale production. According to Hunt's statistics, whereas the average annual output of a Lancashire colliery was 27,200 tons in 1854, it was 36,000 tons by 1871.[14]

Besides the increased cost of new equipment, prosperity in the industry pushed up the value of existing mines. A newspaper article in 1874 stated:

> 'During the last year large fortunes have been made in coals. Collieries worth from £50,000 to £100,000 two or three years ago have been bought by capitalists and companies for £150,000 and £400,000.'[15]

With profits to be garnered at from five to eight shillings a ton, the coal industry was obviously a tempting sphere for speculation. As well as big business, it attracted the small man, hopeful of making easy money. Interest began to be taken in outcroppings on hillsides, especially in North Lancashire, and some millowners hoped to cut out the middlemen by opening their own mines. The great and overriding concern of all of them, big and small alike, was to maximize profit—and this meant maximizing output. To achieve this end, various methods could be used to supplement the increased use of machinery. Priority could be given to getting coal close to the pit mouth and to mining thicker rather than thinner seams. The labour force could be expanded, with daywagemen becoming pitmen. And, finally, they could cut down on the amount of 'dead' work done, i.e. the maintenance of airways, proppings, etc., work not directly related to output.

Dickinson told the Select Committee all four of these means were extensively employed in Lancashire.[16] But some of his remarks aroused the coalowners' anger. They resented his suggestion that they were out to grab the biggest short-term profits regardless of the long-term consequences. Hewlett said that 'to work coalmines unskilfully ... is the most short-sighted policy'.[17] It is probably true that the large companies, like Wigan Coal and Iron, resisted the urge to economize on 'dead' work, but smaller concerns were more likely to dwell on short-run advantages of such moves. North Lancashire's failure to conform to the national pattern was almost certainly because work on ventilation and roadways was skimped so that there would be greater production. Traditionally, it was a district where safety was sacrificed to output.[18]

Ironically, it was just in this period when coalowners had the greatest monetary incentive to increase output that the declining productivity per man became the most pronounced. (See Table 9.) Whereas the annual output per man had been 321 tons in 1870, it had declined to 299 tons in 1872. In part, this was because of the natural tendency for coal to become more difficult to mine. With surface coal being exhausted, colliers were compelled to go deeper. Consequently, the ratio of 'dead' work to actual digging inevitably increased. Another reason was that owners, in a desperate attempt to increase production, hired more workers. Often these were older men or men unaccustomed to mining. As a result, output per man fell.

But, undoubtedly, the main reason was that labour in the coal industry had a backward sloping supply curve. When miners received higher wages, it was not an incentive to work more but to work less. They were able to maintain their modest standards with less time spent underground, and that was what they did. The prosperity of the early eighteen-seventies meant for the colliers a period of greater absenteeism. Coalowners discovered to their chagrin that output was relatively inelastic. The figures given by Hewlett in Table 10, where output declined as coal prices rose, may not be typical of the industry as a whole. Nevertheless, they serve to indicate that the forces of supply and demand work differently with coal to most other commodities.

Naturally, owners blamed the colliers and their union for output failing to rise commensurately with inputs. Their case against the Amalgamated Association of Miners could be summarized under three headings:

First, was the union's militant strike policy. This disrupted production and also served to push up labour costs.

Second, there were restrictive practices, which the union refused to eliminate. The most crippling in Lancashire was the custom of taking tubs from colliers in strict rotation. In other words, a miner who worked quickly could not hew more than his counterpart who worked slowly. This practice helped the aged or less physically robust colliers. It also stripped owners of an important weapon against the union by nullifying the effect of placing a union militant either further from the shaft or where the coal was more difficult to dig. Halliday told the Select Committee it helped the men to counter the employers' discrimination, 'a sort of favouritism in a mine; the man nearest the shaft gets the best chance; and that system was to correct that abuse'.[19]

Third, there was the Amalgamated Association's refusal to countenance the introduction of two-shift working. It had been introduced in the North-East coalfields, where miners belonged to the M.N.A.; Alexander MacDonald favoured its introduction; Thomas Halliday did not. He said, 'It is an attack on the happiness and comfort of the men.'[20] Working late, colliers would find their family life interfered with and they would, moreover, lose the opportunity to attend the Mechanics' Institute.

The effect on coalowners was calamitous. George Elliott told the Select Committee that he had two collieries of the same size in Durham and South Wales. One produced 2,000 tons to the other's 1,000 tons. He attributed this to the Welshmen's aversion to double-shift working. In his Glamorgan colliery, he had the expense of pumping and maintenance for a mere 50 hours labour a week. Elliott said he looked to union leaders of MacDonald's type to overcome the Welshmen's prejudice.[20]

The statistics contained in the Select Committee's report tended to confirm this coalowner's contention. Output per man was very much lower in South Wales than in Durham. In 1872, it was 265 tons in South Wales as against 385 tons in Durham. (See chart.) And, at least, a contributory factor in causing this discrepancy was that the Amalgamated Association men were more zealous than the National Association men in protecting what, to the owners, were restrictive practices.

Of course, ideally, the coalmasters would have liked to have crushed all trade unions. This was not a practical proposition, however. In a period of prosperity, such a move would create widespread chaos, disrupting production and dislocating profits. Some companies had painfully discovered how costly even a short fight against the unions could be. In six weeks, during the 1873 South Wales strike, the Llynvi Iron and Coal Company lost £400,000.[21] So, naturally, most colliery owners sought to avoid such encounters. But the position drastically altered when coal prices—and hence profits—started to plunge. Then came the time to settle old scores, and the owners' main target was the Amalgamated Association.

Internal tensions existed within the Amalgamated Association before the employers made their onslaught in 1874-5. The effect of the depression in the coal trade was that it magnified these tendencies towards fragmentation and disintegration. What it had been possible to cement over whilst the union was winning victories became sources of irreconcilable conflict.

By its very nature, the Amalgamated Association was always subject to serious strains, the difficulty of harmonizing local with national interests. It had been a serious problem to get colliers to adopt a wider view even when things were improving. As we have seen, some districts did not see why they should exercise wage restraint, forego applying for a wage increase, just so all the union's resources could be concentrated on miners elsewhere. When conditions were worsening, the problem became correspondingly greater. It meant asking miners to accept wage cuts—and, in many instances, pay very high levies as well—so that struggles could be put up in other districts where the reductions were even more. This they were extremely loath to do.

During the downswing, events had a momentum of their own. The union executive found itself resisting the most outrageous spate of reductions, only to discover once it had made its decision that much greater wage cuts were to be imposed in another coalfield. Then it had either to divert its resources, giving the impression to the average collier of incompetence, or continue to help miners in one coalfield to resist while, at the same time, telling other miners, faced with a larger cut, to capitulate. Another complicating factor was the astute coalowner who, having calculated that union funds had been earmarked for elsewhere and that no opposition would be made to any cut he might introduce, therefore decided to make it a larger figure than otherwise would have been the case.

In such circumstances, the executive's authority was sure to be questioned. What made its position worse was that it was narrowly based. Its headquarters were at 75 Piccadilly, Manchester, and most of its members were Lancastrians. The reason for this was that quick decisions often had to be made. Hence it was necessary to have men who were on call. Talented leaders from other coalfields usually did not sit on the executive. To do so would have weakened their ties with their own area. They stayed in their respective coalfields, retaining their influence at pit-level, effectively gave the Amalgamated Association a two-tier leadership. Instead of the executive having undisputed power, a pre-requisite if it were to implement a centralized strike strategy, it increasingly found its decisions disputed and disobeyed.

Opposition came in many forms. One of the earliest to appear was the desire to form small local unions. The executive tried to resist this fragmentation. It argued that such disunity weakened the organization, increased administrative cost, and did not afford members the fullest protection. To some extent, the executive's case was undermined because of the position in South Wales. It had been generally agreed that because of the geographical divisions of South Wales, the deep valleys which made communication difficult and unified growth impossible, small local unions, under the aegis of the Amalgamated Association, were the most appropriate form of organization. A report to the Amalgamated Association's Walsall conference stated, 'Small districts were the most workable in South Wales, and the most calculated to increase the members.'[22] Yet, it was a system with disadvantages. The districts differed in their degree of militancy, their strength and their opportunities. Inevitably, there were great variations in wages, which must have been profoundly disturbing to the executive wanting to formulate a national policy of equalization.

Moreover, the existence of small districts in South Wales remained an open incitement to colliers in other coalfields to do likewise. At the St Helens conference in 1872 a lot of time was devoted to discussing cessions. Cannock, with 1,097 members, was granted permission to leave North Staffordshire and Neath was allowed to leave Aberavon. Even Blaenavon, with only 50 members, was permitted by conference, 21 votes to 15, to become a separate district. The conference report, however, states that 'The secretary intimated that Tredegar, with six members, was behind with its subscriptions, and could hardly be considered a district.'[23]

Lancashire colliers, with their strong tradition of local separatism, did not possess the geographical justification of South Wales for creating a proliferation of district organizations. Nevertheless, this did not stop it from occurring. The Wigan district lost Tyldesley in May 1871; then, in 1874, Skelmersdale quit; and, finally, Standish left in 1875. Oldham lost Hollinwood. It would seem that at least two of these district unions—Tyldesley and Hollinwood—were not economically viable and soon ceased to exist. This experience might have provided the reason why the Amalgamated Association forbade Skelmersdale to continue its separate existence. But the Skelmersdale colliers replied that they were tired of the time and money spent in sending delegates to Wigan. Disobeying the conference decision, Skelmersdale continued as an entirely independent entity. From the Amalgamated Association's standpoint, all they had done was to alienate a group of colliers and lessen the amount of subscriptions coming into the national coffers.

The national leadership could not impose too tough a discipline on its affiliated bodies. Were it to do so, then the probability was that they would leave the Amalgamated Association and join its rival the National Association. When the Bristol conference (October 1873) refused to allow Oldham to forget about payment of £290 for special levies still owing, the Oldham district just left the Amalgamated Association and joined the National Association. Likewise, parts of North Lancashire joined with Cumberland and affiliated to the National Association.

Insubordination, a canker that threatened to destroy the Amalgamated Association's existence, was not simply confined to refusal to pay special levies. It extended to the union's whole industrial strategy. By 1874, the executive was expelling colliers, such as those at Cyfarthfa, in South Wales, for the unauthorized acceptance of wage cuts. At the same time, the executive were rebuking others, like the miners of Kingswood, Bristol, 'who, the committee regretted to learn, were in the habit of leaving their work suddenly, without proper notice, for frivolous reasons'.[24]

The decline of the Amalgamated Association began as a result of the heavy drain on its resources imposed by the struggle in Burnley. Traditionally, North Lancashire had been a low-wage, non-union area. A plentiful supply of agricultural labour, eager to work in poor mining conditions because agricultural conditions were even poorer, helped to maintain a position in which the coalowners had almost complete mastery over the men.

A challenge to 'the King Coal tyranny', as the men sometimes called it, came in July 1873, when the Amalgamated Association began a recruitment drive in North Lancashire. Immediately, the executors of J. Hargreaves' collieries dismissed men belonging to the union. Before long, all 1,500 of the firm's employees were in the union and out of work. A strike meeting, held at the Cattle Market, Burnley, on 5 August, was told by Halliday that, 'as God was in heaven these pits would never be worked again until union men were employed and weighing machines on the bank'.[25]

The executors of J. Hargreaves, however, had other plans. They sent sixteen canvassers to Cornwall to recruit an alternative labour force. Despite determined attempts to deter the Cornishmen, the union's own representatives were unable to stop a steady stream of men reaching Lancashire. By November, the employers regarded the struggle as virtually won. They had their full complement of men and output averaged only a thousand to two thousand tons below the ten thousand tons that had been the weekly output before the lock-out began.

In retaliation, the union decided to use the same tactic. If the employers gave the Cornishmen a financial inducement to leave their native county, then the union would give the Cornishmen a financial inducement to return to their native county. A leaflet, addressed to the Cornishmen, promised: 'Miners, on coming out, will be liberally treated.' A meeting was arranged, at the Temperance Hall, Burnley, on 1 December, when Halliday outlined the union's proposals and ended by saying, 'As soon as you desire to go, we will at once make arrangements for your going back to Cornwall.'

The executors of J. Hargreaves soon felt the effects of this union move. Two of their eleven pits had to close. Others were undermanned. And output dropped considerably. They decided to prosecute Halliday and others for conspiracy:

> 'that on 25th November last the defendants did conspire, confederate and agree by divers subtle means to induce and persuade certain miners, workmen and artificers, who had contracted to work for the executors of J. Hargreaves to absent themselves from work.'

Faced with this writ, issued early in January 1874, the Amalgamated Association remained undaunted. At a Burnley meeting in February, Halliday said the union was prepared to fight for a further twelve months if necessary. He also announced that the Amalgamated Association were suing the executors of J. Hargreaves for malicious prosecution.

The cases came before the courts on 18 March.[26] The union engaged three barristers—Aspinwall, Q.C., Hopwood, Q.C., M.P., and Torr, Q.C.—to state its position. They appear to have based their defence on the Amalgamated Association on two main points. First, that the union's officials were honest, law-abiding citizens. In an attempt to prove this, they called many distinguished witnesses who vouched for the character and integrity of the union's officials, especially Thomas Halliday. These witnesses included: the Reverend Dr Jenkins, vicar of Aberdare, H. Lascelles Carr, manager of the

Western Daily Mail, Matthew Bennett, a coalowner, Richard Shaw, M.P., A. Brogden, M.P., a colliery proprietor, and Handel Cossain, a Bristol coalowner.

One can only speculate why so many coalowners were prepared to testify on Halliday's behalf. It may well be that the Amalgamated Association's industrial strategy, which involved reaching mutually binding agreements with coalowners in most sectors of the industry while fighting it out in the remainder, gave the majority of employers greater security than they previously possessed. They knew the union would back them strongly against miners violating an agreement. The union did whatever it could to prevent unofficial strikes. This might have been why Cossain, the Bristol coalowner, testified; the Amalgamated Association condemned the colliers of Kingswood, Bristol, for leaving work 'for frivolous reasons'. On the other hand, coalowners may have been motivated by prudence, a desire not to offend the union, which still remained a mighty giant. A Brogden, M.P., was chairman of the Llynvi Iron and Coal Company that had lost £400,000 unsuccessfully resisting the Amalgamated Association twelve months previously.

But, whatever, their reasons for testifying, the distinguished witnesses repeated that Halliday was 'a man of peace and order' and this could not fail to impress the court. Likewise, the court must have also been impressed with the union lawyers' second point, that in contracts there should be reciprocity. If the employer could sue for breach of contract when the men failed to work, then the men should be equally entitled to sue the employer when he failed to provide work. The executors of J. Hargreaves had not attempted to conceal the fact that they would dismiss most of the Cornishmen once the dispute had ended.

The jury could not agree on its verdict. So the case against the union was dismissed.

Litigation must have been extremely costly for the Amalgamated Association. At the same time, it had to provide strike pay for the Burnley men locked-out. Ordinarily, this would have been ten shillings per man plus a shilling for each dependant child. However, because the union attached such importance to winning the Burnley struggle, it was increased to 15s. 0d. for a man and 1s. 7d. for each child. This was the rate that was paid for nine months. But, in April 1874, the strain on the union coffers was so great that it had to be reduced to 10s. 0d. The decision provoked severe disappointment, and was an important contributory factor to the collapse of the Amalgamated Association's effort at Burnley. In May 1874, those men who were able signed two-year contracts to work for the executors of J. Hargreaves, handing over their union membership cards as part of the contract.

The beginning of the downswing of the trade cycle early in 1874 was the cause of the Amalgamated Association's growing financial plight. Wage cuts were negotiated almost everywhere. But in some places coalowners saw their opportunity to get even with the union. Fisher-Smith, agent for the Earl of Dudley, refused to negotiate with Amalgamated Association officials. An angry encounter ended with him saying he intended to smash the union even if it meant closing the collieries for a year. Miners' leaders left the meeting vowing all the Amalgamated Association's resources would be concentrated on South Staffordshire 'to teach Fisher and the smaller fry who run in the same pool.'[27]

But the Amalgamated Association was not well placed to conduct a struggle in the Midlands. Many of the men there were members of the rival National Association. Enmity had grown up over the years. What was more, the cost of financing the dispute ran at between £11,000 and £12,000 a fortnight.[28] With wage cuts continuing to be made elsewhere, it became increasingly difficult to collect the special levies. In May 1874, Halliday announced the purse was almost empty. Soon after that the headquarters were forced to suspend payments. The Staffordshire miners received strike pay for 12 or 14 weeks of their 17-week strike.

With the funds almost gone, Halliday sought to avoid all industrial conflict. Increasingly, he placed the emphasis on the need for class co-operation. Where owners proposed wage cuts, Halliday counselled their acceptance.

This led to union leaders becoming more and more unpopular. When they advised Wigan colliers to accept a second wage reduction in 1874, Halliday said that 'the men called them all kinds of harsh names'.[29] William Pickard complained that he had 'received anonymous letters from miners in the district as if he were the very dirt upon the street'.[30] As a consequence, they were both badly placed to argue for acceptance of a third reduction in October 1874.

Halliday's tactic was first to get the Wigan district committee in favour of a compromise and opposed to a strike. He then arranged to address an open-air public meeting, where he gave his reasons for accepting the cut: the bad state of trade, the union's depleted finances, the low proportion of men who were union members—4,100 out of 17,000. In his opinion, to fight would only ultimately result in worse terms. Nevertheless, many men argued against him. They said they were still prepared to fight, even if it did lead to defeat, because the wage cut should be opposed. The discussion dragged on until it was dark, and eventually a vote was taken. Amid uproar, the chairman announced that acceptance had been carried. The meeting broke up in disorder with many disputing the chairman's judgment. An impromptu meeting of the Wigan district delegates was quickly convened round a cart. It was resolved to condemn the chairman's sharp practice, publicly repudiate the vote and hold a pit-set conference to take the final decision. Each pit in the district sent two delegates, one to represent the union members, the other the non-union members, to the meeting on 30 September, and the vote there was as follows:

Six for accepting the 10 per cent cut.

Eleven for accepting the 10 per cent cut and asking for an increase.

Eighty-one for no reduction except through arbitration.

As a result, Maskell William Pease, secretary of the coalowners' association, received two conflicting letters. The first was from Pickard and Halliday, saying that the miners accepted the ten per cent cut. The other was from Peter Livesey, general secretary of the Wigan miners, saying that the miners did not accept the ten per cent cut.

Both Wigan and St Helens district unions passed votes of censure on Halliday and Pickard. Through the columns of the press, the Wigan miners challenged Halliday: 'We ask Mr. Halliday what authority he had to sign a paper saying that the men were willing to accept a ten per cent reduction.'[31]

Inevitably, the debate continued at the Amalgamated Association's half-yearly conference at Birmingham, which opened on 8 October. H. Houghton, of Wigan, besides contesting the chairman's ruling over the vote at the open-air meeting, argued that there were many ironworkers, shopkeepers and factory girls present in the crowd that day. Many of them had an economic interest in opposing the strike. Halliday replied that the ultimate decision did not rest with a public meeting or even a district union: it was for the Amalgamated Association's executive to determine strike policy. Robert Lewis, of St Helens, thought that Halliday had manipulated both the open-air meeting and the executive to serve his own ends: 'The men would rather fight and be beaten and have a twenty per cent reduction than accept a ten per cent dishonestly.'[32]

Delegates must have been mindful of the parlous state of the organization. Seven districts had left in the past six months. No financial statement was published, although the *Manchester Guardian* claimed the union had a deficit of £18,500. In these circumstances, and with the remnants of the South Staffordshire district bleating that they should receive their arrears of strike pay, it is hardly surprising that conference turned down, by 34 to 14, Wigan's call for assistance.

A deputation was sent from the Birmingham conference to Wigan to apraise colliers there of the position nationally and to recommend they return to work. The strike had been going for a fortnight, and the deputation appears to have had no effect. A pit-set conference on 12 October, attended by 120 delegates, unanimously re-affirmed the strike decision. A local newspaper commented:

> "The action of the Miners' Conference at Birmingham has greatly incensed the men against the Amalgamated Association and particularly against its president, Mr. Halliday, who they accuse of acting treacherously towards them.'[33]

Dependent entirely upon their own resources, Wigan and St Helens strikers received money from the district union and one shilling and sixpence a week levied on the 1,500 men not involved in the dispute. But these funds were meagre. Soon they were compelled to resort to traditional methods, such as begging. The general public, recalling the prosperity of the coal trade only a few months previously, were not so responsive:

> 'We are met with the cold shoulder on every side, and with the insulting expression—"What, colliers' wives begging already? Where is the ten, twelve and twenty shillings a day you've been getting? Go away, I shall give you nowt." '[34]

Some then turned to stealing as a stratagem. The pilfering of vegetables was widespread in country districts. At Burtonwood six geese were taken and a gander left with a piece of paper, containing sixpence, round its neck:

> 'We've travelled far all day
> And further will not wander.
> We've bought the geese for a penny-a-piece,
> And left the money with the gander.'

Bitterness entered the dispute. A Standish grocer, Ryder, who refused to give colliers credit, had his hay-rick burnt one night. Bottles of gunpowder were thrown through the windows of houses of blacklegs at Haydock. A serious disturbance occurred outside the Ram's Head, Haydock, on 23 October when 500 strikers met 200 strikebreakers returning from work.

The strike ended in a more ragged way than most strikes. The majority of men were compelled to accept a ten per cent cut. But others, especially in the St Helens area, where the price of coal had risen, gained a ten per cent rise. Nevertheless, however, they fared, colliers appear to have reached the same conclusion, namely, that, as they had received no help from the Amalgamated Association, then it was a useless organization. As one Wigan miner put it: 'let the rotten ship of the Amalgamation sink into oblivion for it is not fit to carry the mining population of Great Britain in a storm'.[35]

The plight of the Amalgamated Association worsened. In January 1875, South Wales coalowners decided to impose a wage cut. The workers replied by saying they would only accept a cut if it were the result of an arbitration decision. They also called upon the owners to open the books to trade union inspection since they questioned the need for a cut. Within a short time, 50,000 men in South Wales were idle. This placed a heavy strain on the union's slender resources. Halliday admitted it faced bankruptcy and he strived strenuously to achieve a settlement. But the South Wales coalowners now seized their chance. Most of them had never liked Halliday; now they refused to negotiate with him. In vain he appealed to Lord Aberdare, formerly H. A. Bruce, the Home Secretary, who was himself a Welsh coalowner, to help in obtaining a settlement. But His Lordship declined avvering that, 'while there was a superfluity of labour, employers were justly entitled to the full advantage of it, irrespective of price or profit'.[36]

The South Wales owners stated that they were only prepared to have discussions with representatives of the more moderate union, the National Association. So two of its officials, Crawford and Casey, journeyed to Swansea. But the terms offered appear to have been unacceptable. For the following week MacDonald did something which, for him was unprecedented and against the rules of his union—he issued an appeal for strike funds:

> 'The miners of Scotland and the miners of England: unless they are willing to have the slave law placed on them for many years to come, should rally round the men of South Wales. From every miner in both countries should be sent not less than a shilling a week to aid the men now out, that is, if they love liberty.'

The appeal had a much more militant tone than was customary for MacDonald. After saying that the coal industry of South Wales was dominated by four large families—the Crawshays, Fothergills, the Vivians and Davies—he continued:

'The wealth of a few great families had been collected from the workmen and produced by the workmen; and he questioned the right of men to turn round with their wealth and try to starve the workmen, their families and their children.'[37]

MacDonald's appeal, like that of Halliday, met with a poor response. Miners, themselves suffering hardship, were reluctant to impoverish themselves further. Only £1,800 came from the Amalgamated Association's English branches.[38] Moreover, the situation was worsened because, in an effort to retain unity among strikers, the money was evenly distributed between unionists and non-unionists alike. But even this tactic did not avert disaster. When, after the dispute had lasted almost four months, the owners offered to cut wages by 15 per cent, many non-members found it too tempting to reject. A trickle back to work soon became a deluge—the strike was over.

The collapse in South Wales meant the final bastion of the Amalgamated Association had fallen. Halliday's position had also been undermined in a number of ways; the failure of the owners to negotiate with him; the encroachment on his territory of the MacDonald union; the emergence of William Abraham—'Mabon'—as a moderate leader of great stature, a man with more influence than Halliday in South Wales.[39]

A last desperate attempt was made in April 1875, at the Swansea conference, to reorganize the battered union. But it was quite useless. So, when approaches were made by the National Association for unity of the two organizations, then the Amalgamated Association leaders grasped the opportunity with alacrity. They attended a joint meeting at Leeds Town Hall on 26 April, where it was agreed that the new body would be run on federal—i.e. MacDonaldite—lines.[40] The idea of centralized control was completely rejected.

All that remained to be done was to wind up the Amalgamated Association. The last conference was held at the Corn Exchange, Shrewsbury, on 24 August, 1875, and conference was unanimous that it should join the new organization, to be known as the Miners' National Union.[41]

The failure of the Amalgamated Association can be attributed to mounting opposition from miners to its tactics. They resented subordinating local interest to an overall plan, to paying large sums in levies, and to abiding by unpopular arbitration decisions. This growing criticism expressed itself through the decline in membership and by desire for separatism. At a different level, the union can be regarded as a victim of the business cycle; it gained more than its fair share of praise for pushing wages during the upswing; it had more than its fair share of blame for the reduction of wages during the downswing. Miners failed to appreciate to the fullest extent, the effects of the working of market forces, which the Amalgamated Association could only influence in a marginal way.

Table 9

Labour Force in British Coal Mines

	No. of Miners	*Output per man (tons)*
1864	307,000	309
1865	315,000	313
1866	321,000	314
1867	333,000	316
1868	347,000	302
1869	345,000	316
1870	351,000	321
1871	371,000	317
1872	413,000	299

Source: Select Committee's report on the Scarcity and Dearness of Coal, 1873.

Table 10

Wigan Coal and Iron Company, 1867-1873

	Output of Coal	*Price of Coal*
1867	100	100
1868	82·4	93·5
1869	86·9	91
1870	95·9	91
1871	98·2	92
1872	95·1	134
1873	73·0	182·5

Source: A. Hewlett's testimony to Select Committee on the Scarcity and Dearness of Coal, 1873.

The figures for 1873 were based on the first three months.

A disaster at Blundells' colliery, Wigan, in 1877. These pictures are taken from the *London Illustrated News.*

'Surveying under difficulties' is the caption given to this drawing by *The Graphic*, of 21 September, 1878, to illustrate an article on working in a coal mine.

Table 11

Fairbottom Colliery in Oldham

	Output	*No. of Men*	*Output per Man*
1867	100	100	100
1868	96·1	94·4	101·9
1869	86·7	78·2	110·7
1870	96·5	82·7	116·8
1871	111·0	108·9	110·0
1872	111·8	129·9	86·1
1873	121·8	129·9	93·9

Source: Isaiah Booth's testimony to the Select Committee on the Scarcity and Dearness of Coal, 1873.

Table 12

Astley and Tyldesley Coal and Salt Company
Statements of Production Costs and Output, 1871-6

	Output	*Colliers' Wages*		*Daywagemen*	
	(tons)	*Coals*	*Yard*	*Underground*	*Bank*
uly-Dec. 1871	81,785	£6,442 (1/6$\frac{7}{8}$)	£1,406 (4$\frac{1}{8}$)	£3,188 (9$\frac{5}{8}$)	£2,368 (7)
an.-June 1872	82,233	£7,158 (1/8$\frac{7}{8}$)	£1,224 (3$\frac{9}{16}$)	£3,469 (10$\frac{1}{8}$)	£2,458 (7$\frac{3}{8}$)
uly-Dec. 1872	82,104	£9,000 (2/2$\frac{5}{16}$)	£1,540 (4$\frac{1}{2}$)	£4,646 (1/1$\frac{9}{16}$)	£3,012 (8$\frac{13}{16}$)
an.-June 1873	87,089	£10,606 (2/5$\frac{1}{4}$)	£2,058 (5$\frac{5}{8}$)	£6,577 (1/6$\frac{1}{8}$)	£3,527 (9$\frac{3}{4}$)
an.-June 1874	84,703	£9,956 (2/4$\frac{3}{16}$)	£2,277 (6$\frac{7}{16}$)	£6,430 (1/6$\frac{1}{4}$)	£3,715 (10$\frac{1}{2}$)
uly-Dec. 1874	74,447	£7,687 (2/0$\frac{3}{4}$)	£1,780 (5$\frac{11}{16}$)	£6,292 (1/8$\frac{1}{4}$)	£3,720 (1/0)
an.-June 1875	99,318	£10,297 (2/0$\frac{7}{8}$)	£1,972 (4$\frac{3}{4}$)	£7,591 (1/6$\frac{3}{8}$)	£4,033 (9$\frac{3}{4}$)
uly-Dec. 1875	108,631	£10,693 (1/11$\frac{5}{8}$)	£2,571 (5$\frac{11}{16}$)	£8,281 (1/6$\frac{5}{16}$)	£4,107 (9$\frac{3}{16}$)
an.-June 1876	113,381	£11,169 (1/11$\frac{5}{8}$)	£2,623 (5$\frac{5}{8}$)	£7,861 (1/4$\frac{5}{8}$)	£4,317 (9$\frac{1}{8}$)

	Wage Cost per ton	*Total Cost per ton*
July-Dec. 1871	3/3$\frac{3}{8}$	5/4$\frac{1}{8}$
Jan.-June 1872	3/5$\frac{3}{4}$	6/0$\frac{1}{2}$
July-Dec. 1872	4/5$\frac{3}{16}$	7/9$\frac{3}{16}$
Jan.-June 1873	5/2$\frac{3}{4}$	9/1$\frac{7}{8}$
Jan.-June 1874	5/3$\frac{3}{8}$	8/10$\frac{7}{16}$
July-Dec. 1874	5/2$\frac{11}{16}$	8/3$\frac{3}{4}$
Jan.-June 1875	4/9$\frac{3}{4}$	7/7$\frac{7}{8}$
July-Dec. 1875	4/8$\frac{3}{16}$	7/0$\frac{9}{16}$
Jan.-June 1876	4/7	7/0$\frac{1}{4}$

Source: Lancashire Public Records Office. Balance-sheet for July-December 1873 missing. Figures in brackets—cost per ton.

CONFERENCES OF THE AMALGAMATED ASSOCIATION

Preliminary conference, Manchester, July 1869.
Decisions: (i) to form the union.
(ii) to agitate for higher pay.
Membership: 6,500.
Comment: Confined entirely to Lancashire and Cheshire miners.

First conference, Manchester, August 1869.
Decisions: (i) to formulate rules of the Amalgamated Association.
(ii) to accept a centralized wage strategy (Rule 13).
(iii) to campaign to recruit other coalfields.
Comment: Membership still confined to Lancashire and Cheshire.

Second conference, Bolton, October 1869.
Decisions: (i) to contact Miners' National Association for joint activity or unity.
(ii) to discuss the question of shorter hours at next conference.
(iii) to continue agitation in other coalfields.
Comment: Seven delegates from North Wales attended, the first from any coalfield outside Lancashire.

Third conference, Wigan, January 1870
Decisions: (i) to make restriction, not strike action, the tactic in Farnworth, where no wage increase had been obtained.
Membership: 12,000.
Comment: Forty delegates attended. Mood of conference more confident than hitherto, the result of the partially successful wage campaign.

Fourth conference, Wrexham, April 1870.
Decisions: (i) to make no further approaches to the Miners' National Association.
(ii) to leave the wage strategy to the executive. This decision, tantamount to re-affirming constitution, only carried 17 votes to 15.
(iii) to appoint miners' agents at the conference and for them to hold their appointments from one conference to the next.
Comment: Meetings held throughout North Wales in an attempt to strengthen union organization there.

Fifth conference, Hanley, October 1870.
Decisions: (i) to re-affirm decision to leave strike strategy to the executive carried 18 votes to 9.
(ii) to refer an appeal for financial aid from West Yorkshire to the districts.
Comment: Delegates from South Wales as well as North Wales, Staffordshire and Lancashire. Also, West Yorkshire made feelers about joining, probably because it wanted help in its disputes. And Northumberland was contemplating joining the Amalgamated Association. Halliday hoped the Amalgamated Association would become truly national, not by uniting with the Miners' National Association but by supplanting it.

Sixth conference, Manchester, May 1871.

Decisions: (i) to attempt to get a wage increase, first in South Wales, then in North Staffordshire, and finally for the entire membership. Detailed tactics left to the executive.

(ii) to instruct Tyldesley to rejoin the Wigan district.

Membership: 23,491, of which 14,000 were in South Wales.

Comment: Halliday spoke of the continuing lack of confidence between English and Welsh colliers. There was probably some truth in this.

Seventh conference, Merthyr, October 1871.

Decisions: (i) to condemn the Criminal Law Amendment Act (1871) as 'a sinister invasion of the rights of Englishmen'.

(ii) to approach to Miners' National Association for a joint campaign on the question of mining legislation.

(iii) to discuss within the districts the desirability of setting up a parliamentary fund.

Comment: Conference probably held in Merthyr to celebrate the victory over the South Wales steam coalowners. But success had been won at a price: only £200 remained in the union coffers after paying conference expenses.

Eighth conference, St Helens, April 1872.

Decisions: (i) to permit Cannock (1,097 members) to cede from North Staffordshire (5,989 members). But no decision on how to divide the district's funds.

(ii) to pay union officers eight shillings a day plus second-class rail fares.

(iii) to set up a parliamentary fund to pay for members' election expenses.

Membership: 23,676.

Comment: In the past year membership had remained virtually static, although considerable wage increases had been won. Total income for the year, excluding special levies, was £1,966, hardly a large sum.

Ninth conference, Walsall, October 1872.

Decisions: (i) to thank politicians who had helped the union during discussions on the Mines Regulation Act (1872).

(ii) to reduce the executive from nine to seven.

(iii) to re-affirm decision to have a political levy and to ask other trades to co-operate.

(iv) to allow enginemen, tippers and blacksmiths to join the union, although Pickard wanted them included.

Membership: 70,536.

Comment: The Amalgamated Association had now become extremely powerful. This fact is reflected in the amount of newspaper comment on its conference decisions.

Tenth conference, Newport, Monmouthshire, April 1873.

Decisions: (i) to condemn the colliers at Rhondda Merthyr Steam Coal Company for breaking the agreement reached with the employers after the 1871 dispute, Halliday sent to remonstrate with them.

(ii) to investigate the possibility of starting a new colliery to be run along co-operative lines.
(iii) to refuse South Staffordshire's request to be allowed to put in for a wage increase since it conflicted with previous decision to leave the question of wages to the executive.
(iv) to put forward Halliday (Merthyr) and Pickard (Wigan) as parliamentary candidates.

Membership: 86,623.

Comment: There had been twelve levies for the South Wales strike, but this had not stopped membership from growing. In the past six months contributions had been £5,855 while special levies had been £43,831.

Eleventh conference, Bristol, October 1873.

Decisions: (i) to call upon all districts of the union to investigate the possibility of putting forward parliamentary candidates.
(ii) to give £1,000 to the Plimsoll Fund and £1,000 to the Labour Representation League.
(iii) to consider the possibility of the union starting its own school to train miners to be managers.

Membership: 99,399.

Comment: Although membership was still growing, there were ominous signs: (i) district falling behind with payments of special levies—Oldham owed £290 and North Wales £141; (ii) parts of North Lancashire joining with Cumberland and joining the Miners' National Association; (iii) the beginning of the protracted and costly dispute at Burnley.

Twelfth conference, Manchester, April 1874.

Decisions: (i) to ask for the inadequacies in the Mines Regulation Act (1872) to be remedied.
(ii) to put forward further candidates in future parliamentary elections and to have another threepenny levy towards that end.
(iii) to ask the Shropshire colliers, who were out on strike, to resume work since they had not first sought permission from the executive.

Membership: 106,000.

Comment: Membership had reached its peak. Nevertheless, the downward pull of wages had begun. Oldham (membership 1,200) had resigned from the Amalgamated Association and joined the Miners' National Association. Total income had started to decline.

Thirteenth conference, Birmingham, October 1874.

Decisions: (i) Wigan instructed, by 34 votes to 14, to accept a wage cut.
(ii) Resolutions passed in favour of co-operative collieries, technical education and sanitary reform.

Membership: 57,766.

Comment: The Amalgamated Association in real difficulty. Midlands delegates opposed Wigan's plea for the right to resist wage cuts because Staffordshire colliers were still owed strike pay. Seven districts had ceded from the Amalgamated Association and to these were added, after conference decisions had gone against them, Wigan and St Helens.

Fourteenth conference, Swansea, April 1875.

Decisions: (i) to revise the union constitution.
(ii) to send union representatives to North Wales in an attempt to resolve the strike there.
(iii) to appoint Abraham ('Mabon') as vice-chairman in place of William Pickard, whose district, Wigan, had left the union.

Membership: 50,000 claimed. (This is almost certainly an overestimate as many districts were behind with subscriptions.)

Comment: The nine-day conference spent seven days deliberating on revisions to rules. To hold highly detailed, complex constitutional debates while the union was breaking up around them might seem short-sighted and strange but it must be remembered that these discussions were a symptom of the malaise, the pervasive feeling that something was wrong with the organization and something needed doing about it.

Fifteenth conference, Shrewsbury, August 1875.

Decisions: (i) to dissolve the Amalgamated Association.
(ii) to get the rump of the membership to join the Miners' National Union.

Membership: 25,000 claimed—a very questionable claim.

Comment: The Amalgamated Association had virtually ceased to exist before the Shrewsbury conference was held. Failure to be able to provide financial support had led the Lancashire, Staffordshire, and finally the Welsh colliers to leave.

CHAPTER EIGHT

The Miners and Politics (1870-5)

In 1870, the attention of mining communities focused on Parliament. A spate of pit disasters underlined the urgency for improved legislation. Colliers submitted a petition, with 108,000 signatures, and held public meetings to call for immediate action. But Parliament appeared unresponsive, a fact miners ruefully attributed to the 180 M.P.s who were connected, directly or indirectly, with the coal trade. 'If the House of Commons thought as much about the mining community as the cattle plague,' William Pickard told a Wigan rally, 'the measure would have been passed long ago.'[1]

Mounting pressure eventually forced the Liberal Government to react. The Reform Bill of 1867, enfranchising the urban proletariat, necessarily meant that governments, concerned about their own survival, had to take more heed of the opinions of British workers. Yet, in keeping their goodwill, the Liberal Government had been none too successful. It found parliamentary time for the Criminal Law Amendment Act, a measure unpopular with workers because it bristled with dangers for organized labour, while it could not find time for a measure to safeguard miners' lives. Faced with growing criticism, Gladstone appeased the colliers by publicly expressing his regret over the postponement of the Mines Inspection Act. Stating this in a letter to Enoch Robinson, of Duckinfield, he nevertheless stressed that three of the five measured that took the major portion of the last session—the Irish Land Bill, Education Bill and Ballot Act—were 'mainly concerned with the welfare of the labouring population'.[2]

While the Gladstone administration, mindful of its public image, braced itself for the introduction of mining legislation in the 1872 session, many coalowners, particularly the smaller ones, viewed the prospect with misgivings. For the marginal producer, stricter regulations with their ensuing higher costs threatened his existence. Even where such dire consequences were unlikely, the prospect still remained unappealing: higher costs would either eat into profit margins or involve a higher market price. An article in the *Colliery Guardian*, ascribing the colliers' clamour from fresh legislation to 'intense class selfishness', regarded it as a facet of their general policy, to create artificial restrictions and do as little work as possible. The result would be diminished investment, which would have a deleterious effect on the economy: 'They cannot see that cheaper coal means cheaper houses, cheaper clothing, cheaper food.'[3]

Not all coalowners shared this attitude. Some of them, especially the larger producers, took a more far-sighted and dispassionate view. They did not oppose regulations that would make it compulsory for colliery officials to possess proper qualifications. They recalled the recent findings of the Royal Commission on Coal (1871), which stated that waste of coal due to bad working or carelessness could be as much as forty per cent of total output whereas, by the best methods, it could be reduced to ten per cent. Technical experts reduced costs as well as lessening the likelihood of disasters, calamities where damage could reach £30,000.

Therefore, coalowners and miners alike might benefit from many accident prevention measures. More rigorous inspection, the provision of detailed, scientific knowledge, would reduce the number of holocausts, costly in lives and capital. Similarly, the proposal that all underground roadways should be properly propped and spragged required implementation for efficient, not to mention safe, mining. Moreover, if new regulations resulted, in the short run, in higher costs, making it impossible for marginal producers to remain in business, it would mean a bigger share of the market for the rest.

A large coalowner like Lord Elcho, aware of the confluence of interests between capital and labour, sought to establish cordial contact with miners' leaders. He made it plain he supported their campaign for improved legislation. His declaration afforded him public acclaim for his humanitarian and philanthropic impulses. Yet, he could not support them all the way. The point of disagreement was over what sanctions, if any, should the Act impose on defaulters. Lord Elcho could not countenance a bill that would render himself and other coalowners liable to criminal proceedings. But without such penalties, argued the advocates of the bill, the Act would be useless; owners could disobey its clauses with impunity.

Lord Elcho's stand placed Alexander MacDonald, president of the Miners' National Association, in an awkward position. He was liable to offend his followers or lose His Lordship's esteem. Wishing to avoid a clash with either party, he kept quiet about Lord Elcho's move. But this did not prevent the news from getting out.

Professor Beesly wrote to the rival union, the Amalgamated Association of Miners, to warn the colliers what was afoot. His letter, read to the A.A.M. Merthyr Tydvil conference, began by contrasting the lack of regulations in the coal industry with the large number of regulations governing the railways:

> 'Railway companies are tied down by minute regulations and are made to pay heavy sums as compensation for injuries to passengers. But then the upper and middle classes travel by railways. If the gentry and shopkeepers of the mining districts had to pass a few hours occasionally in the pits, we should soon have the needful legislation.
>
> 'When a miner is found smoking his pipe in the pit, he is sent to prison for three months, and right enough ... (But) if the pitman is sent to prison for lighting his pipe on a chance occasion, how much more heavily should the employer be punished who habitually keeps his pit in a dangerous state? Lord Elcho not long ago complained that, under a Bill brought in by Mr. Bruce, an employer might be summoned home from the Continent to undergo imprisonment for an accident in his pit. Why not? Why should a man be exempt from punishment because he can afford to travel on the Continent? The days are coming when we will teach Lord Elcho and others of his species the theory and practice of equality.'[4]

Thomas Halliday, presiding, moved that a letter of thanks be sent to Professor Beesly. Conference followed this up by expressing its hostility to the middle-class Parliament and its intention to seek the return of M.P.s who would promote working-class interests.

In practical terms, the A.A.M. could do very little. Meetings and petitions had an extremely limited impact. More important was to possess influential political friends. Here, the A.A.M. was at a disadvantage when compared to Alexander MacDonald, who, for many years, had sedulously sought to establish contacts in the corridors of power. He was well-acquainted with politicians and parliamentary procedure. As *The Beehive* observed, 'Mr. MacDonald is so much of the year in town, and appears to aim at managing the parliamentary business of unionists generally, as he already does that of his own trade.'[5]

During the discussions that accompanied the passing of the Mines Regulation Act (1872), MacDonald was naturally regarded, by the government and M.P.s as the spokesmen for the miners. It was in this context that he had to decide where he stood, backing Lord Elcho or the miners' petitions.

When MacDonald's attitude became known, the home secretary expressed concern that the miners had changed their minds, *The Times*, in its parliamentary report, stated:

'On the preamble of the Bill being put, Mr. Bruce wished to take that opportunity of doing an act of justice to a number of gentlemen who had acted as delegates from colliers in the north during the discussion of this measure. In consequence of representations which had been made to him he had complained, and felt himself justified in complaining of the course they had taken in regard to the Bill, considering that from all the districts of the country the colliers had sent up petitions to that House, and memorials to his right honourable friend at the head of the Government and himself, praying that the Bill in its entirety should pass. He confessed himself deeply disappointed when he found, with respect to some of the most important clauses of the Bill, of which colliers had expressed their approval, their representatives in London were quoted as opposed to the proposals of the Government, and he drew particular attention to the fact that their consent had been obtained to the insertion of words which, in his opinion, went far to destroy the legislation proposed for the protection of children, and also for the protection of the colliers themselves from accident. He asserted on that occasion that he had been informed that they would not support the Government with respect to the general rules on conditions that they obtained payment by weight instead of measure.'

When the bill came for discussion in committee, Staveley Hill, an M.P. who was a coalowner, moved the insertion of the word 'knowingly' into clause 14, which imposed penalties for the employment of persons contrary to the provisions of the bill. The Home Secretary, Mundella and others opposed the amendment on the grounds that, if adopted, it would make the conviction of offending coalowners virtually impossible. *The Times* endorsed this opinion: 'Mr. Hill's amendment, whether so intended or not, will practically shield deliberate negligence from any penalty.'

One M.P., Roden, supporting Hill's amendment, said, 'He was sorry the Home Secretary did not accept the amendment which had been approved by the representatives of the working miners.' Another coal-owning M.P., Liddell, expressed the hope 'that the Government would assent to the amendment. The men engaged in mines admitted themselves that the insertion of the proposed words was only a matter of justice.' While Lord Elcho, by far the most outspoken, 'regretted that the Home Secretary would not accede to the amendment. He admitted the soundness of the principle that the Government could not consider themselves bound by any decision come to by any body of men out of the House; but when persons who were the most interested in the operation of this Bill agreed upon any point in connexion with it, that might be taken as *prima facie* evidence in its favour. Seeing that the workmen wished the proposed words to be introduced, he would certainly vote for the amendment if it were pressed to a division.'

The Hill amendment was passed. When, however, the coalowners adopted the same tactic on payment by weight—that of stating they had the support of the miners' representatives—the House of Commons defeated it.

The conduct of MacDonald, as leader of the miners' delegation, aroused considerable criticism in trade union and radical circles. Frederick Engels wrote, 'MacDonald, however, betrayed the workers of his own trade when, during the negotiations on the last mining law, which he attended as the representative of his trade, he sanctioned an amendment which was so grossly in the interests of the capitalists that even the government had not dared to include it in the draft'.[6]

But far more damaging than Engels' was Beesly's attack, stretching for over two months in *The Beehive*. Hitherto Beesly had held MacDonald in high regard. When, in 1868, MacDonald had contemplated standing for Parliament at Kilmarnock, Beesly had written an open letter, supporting his candidature and saying: 'You are one of the few representatives of Unionism I know who would be able to fight the battle with effect in such an Assembly as the House of Commons.'[7] Four years later, Beesly's opinion of MacDonald had changed. He now accused him of flouting the interests of the colliers: MacDonald knew what legislation the colliers wanted; he knew before Whitsun what was the Government's attitude to the various amendments; then he supported the coalowners.

When MacDonald replied that the Act contained many admirable clauses, Beesly pointed out:

> 'he overlooks the fact that the insertion of the little words "knowingly" and "under normal circumstances", would render the most perfect Bill worthless. What would be the use of laws prohibiting murder, theft, or forgery, if they contained words which made it almost impossible to enforce the penalties? The Bill might have passed with those amendments, and would very likely have been welcomed by the miners as a triumphant settlement due to the services of Mr. MacDonald. Then would come the disappointing experience; and when it was found that coalowners could not be

convicted, and broke the law with impunity, there would be more agitation, more sending of Mr. MacDonald to London, more conferences, and more championship of Lord Elcho. These pleasant results have been partially secured as it is by the insertion of the word "knowingly".'[8]

Beesly's prediction proved correct. Initially, many colliers were impressed by the fine sentiments of the Mines Regulation Act (1872). In Scotland, they even celebrated its passing in November of that year by presenting MacDonald with a silver casket containing £680 while a further £100 was collected from English miners, mainly in Yorkshire.[9] But soon the inadequacy of the Act became painfully apparent. By 27 June, 1873, 91,000 in thirteen counties were petitioning the home secretary for improved legislation.[10]

In his dispute with Beesly, MacDonald had defended his close association with Lord Elcho: 'Let me only further state what made the Bill of 1872 acceptable to the mining population was the Government adopting all the amendments the noble lord so persistently and so ably advocated.' Beesly, on the other hand, thought:

> 'Mr. MacDonald knows perfectly well that Lord Elcho has not only acted in the interests of the coal-owners on all important points, but that he has uniformly opposed the cause of trade unions, in and out of Parliament, for many years. One need not look any further than the report of the trade union commission for the proof of what I say. It is simply scandalous that any representative of unionists should take such a man into alliance and entrust the business of workmen to him in the House.'[11]

The only arguments for collaboration with Lord Elcho—ones that MacDonald did not use—were, first, given the outlook of the gentlemen in charge of the legal system, it is unlikely that, even if the Act had contained punitive clauses, they would have been applied with vigour. But, second, it can be contended that, were punitive penalties enforced, this might have a bad effect, arousing resentment and hostility among coalowners where co-operation resulting from greater education and a changed climate of opinion was what was required. In the long run, persuasion might be a more powerful force than compulsion.

But, as it was, MacDonald stood virtually defenceless before his critics. In desperation, he wrote a pamphlet on the Mines Regulations Act, extolling its merits.[12] He also wrote to the Home Secretary, imploring him to make a statement that would provide aid and comfort.

However, Bruce declined to assist while, in private correspondence, he expressed grim satisfaction at having 'MacDonald and other delegates on their knees'.[13]

In *The Beehive* Beesly pointedly referred to MacDonald's 'patron, Lord Elcho'.[14] Mundella's verdict was similar: 'a dishonest but clever and unscrupulous miners' agent ... this bad fellow who is Lord Elcho's Limited.'[15] A statement that presumably implies that MacDonald had been bought by His Lordship, who used him as part of his property.

It is impossible to say whether these imputations were malicious or grounded in fact. What is clear, however, is that MacDonald's financial affairs are tinged with mystery. In 1868, one of his principal reasons for withdrawing from the Kilmarnock election was lack of funds.[16] But five years later, when the M.N.A. Leeds conference of 18 November, 1873 discussed the question of parliamentary candidatures, MacDonald stated he needed no financial assistance for his fight at Stafford.[17] Yet, this contest must have been very expensive. At the same 1874 general election Pickard fought Wigan and Halliday fought Merthyr, costing £680 and £960 respectively. The Amalgamated Association had considerable difficulty, despite a levy imposed on all its 98,000 members, in raising money to pay for these two contests. How MacDonald, relying upon his personal wealth and contacts rather than union support, could raise sufficient remains difficult to explain. Having won Stafford, MacDonald does not seem to have been dismayed at the prospect of being an M.P., with no salary, and the expense of living in London. In 1874, he announced at the T.U.C. that he was giving a prize of £50 for the best essay on trade unions.[18]

There appears to have been three conceivable sources of income for MacDonald. First salary from the union, which paid its officers eleven shillings a day while in London and 8s. 3d. in the provinces.[19] From this MacDonald might have derived an annual income approaching £200. But he told the M.N.A. Saltburn-on-Sea conference, on 29 August, 1875, that he had worked for the Association for the past twelve years without drawing his salary.[20] If this statement is true, during that time MacDonald must have had a personal source of finance, either business speculation and/or payment for services rendered to coalowners.

The growing disappointment with parliamentary discussions over the Mines Regulation Act formed the background against which the Amalgamated Association considered the question of sponsoring candidates at the next general election. It did not intend to repeat the mistake made at Kilmarnock in 1863, when the Scottish miners decided to have a voluntary levy of a shilling per head; both the scheme and the candidature fell through.[21] What was required was a compulsory levy of the entire membership, a separate fund for parliamentary purposes, William Wogan, of Wigan, moved that the A.A.M. adopt this procedure at its St Helens conference on 1 April, 1872. By passing his resolution, the Amalgamated Association had created, as Mr W. B. Gwyn points out, 'the first union-wide political levy'.[22] But the A.A.M. did not act hastily. At the Walsall conference, of October 1872, it resolved to approach the M.N.A. for co-operation over the adoption of candidates.

Ironically, the M.N.A. which emphasized the importance of legislation, could not achieve sufficient internal unity to sponsor any candidates. The loose, federal structure permitted personal jealousy and local patriotism to be predominant. While MacDonald wanted the M.N.A. to adopt a scheme similar to the A.A.M., leaders in the various coalfields were not prepared to collect funds to be used in another coalfield. Eventually the M.N.A. resolved that each of its affiliated bodies be allowed to do as they deemed advisable.[23] In practice, this meant that the Northumberland miners, supporting their secretary Thomas Burt, had to go it alone. Besides having to meet his election expenses, they then had to find a parliamentary salary of £350 once he had been elected M.P. for Morpeth, in addition to the £150 Burt still received as their secretary.[24] The Northumbrian miners had no help from elsewhere.

In the same general election MacDonald successfully stood at Stafford. His campaign was run entirely independent from the M.N.A. There were no miners living in the constituency, and his election address did not mention them.[25]

Burt and MacDonald were the first workmen to be returned to Parliament. Probably the most important reason for their victories was that in both constituencies they stood with the blessing of the Liberals. This is an important distinction between them and the two officially sponsored candidates of the Amalgamated Association, Halliday and Pickard, both of whom stood against Liberal candidates.

Throughout the discussions inside his union, Halliday had favoured a line of political independence. He wanted pure working-class representation, without any compromising alliances. His views are expressed in such remarks as 'he would not join any party that sought to sell the interests of the working men in any way. He thought a man ought not to go to Parliament either as a Liberal or Conservative.'[26] On another occasion, he said that 'whatever was against the working men, they found the Tories and Whigs banded together'.[27]

Halliday's opinions caused dismay. The Labour Representation League, an organization whose aim was to promote 'the return to Parliament of qualified working men', believed this object could best be attained through collaboration with middle-class Radicals.[28] The president of the League, R. M. Latham criticized Halliday at a London meeting. He said it was important they did not 'lose to the cause of labour representation the support of every thoughtful liberal politician'. While the L.R.L.'s first aim must always remain to get workmen elected, Latham said:

> 'Our next is to return, or aid in returning, men of other classes who have studied working men's question's and whose views are favourable to the promotion of working men's interests. The latter point is essential. I, for one, repudiate the sentiments represented by the press as having being put forward by Mr. Halliday.'

Underlying the Halliday-Latham dispute may have been rivalry over whom should contest the Wigan constituency. In April 1873, only a few weeks before his death, Latham made it clear that he intended to stand. Had he done so, he would have been a strong candidate. Born at Wigan on 8 July, 1834, he was educated at the local grammar school and Heidelburg University. In 1860 he was called to the bar. Besides being a barrister, he took a lively interest in scientific invention, technical education, and was the founder of a learned journal, the *Scientific and Literary Review*.[29] In his book on 'British Working Class Politics, 1835-1914', Professor Cole states that Latham was the prime mover in getting the warring groups—the Junta, Potter, the middle-class radicals etc.—to sink their differences and form the Labour Representation League.[30] His distinguished record gave him a strong claim to contest Wigan; the miners' counterclaim was based on their numerical strength: of the constituency's 5,062 voters, 3,400 were workers, the largest section of whom—903—happened to be miners.[31]

Latham's sudden death resolved the dispute, leaving the miners free to run William Pickard.

Wigan, a two-member constituency, was at the time represented in Parliament by Liberals, Lancaster and Wood. They had been returned at the general election of 1868, held in the midst of the extremely violent Lancashire colliers' strike. (See chapter four.) As a director of

Wigan Coal and Iron Company, John Lancaster was as much involved in the industrial dispute as he was in the election. He was largely responsible for the mass importation of Staffordshire strikebreakers, for calling up the authorities to use troops to dislodge colliers who had occupied pithead premises. The heavy sentences that followed the acts of violence had his approval. So detested was Lancaster by the strikers that his presence in the town centre sparked off a riot from which he was only able to extricate himself thanks to the possession of a gun. In these circumstances, it seems amazing that a man who was hated so much for his industrial activity should still retain sufficient political popularity to get elected.

The mechanics of this miracle are revealed in the George Howell Collection, at the Bishopgate Institute, London. Howell, himself a trade union leader, endeavoured in the 1868 election to secure the newly enfranchised working class for the Liberals while, at the same time, werever possible securing seats for candidates favourable to his political views. His agents visited sixty-five boroughs, and among the reports he received was this from Wigan:

> 'The Liberal Candidates for this Borough are Messrs. Wood and Lancaster. The first-named gentleman being the sitting member. His election is sure, but as the election of the second gentleman depends largely on the miners, and they, having a serious trade dispute with him, his success is not so certain.
>
> Mr. Lancaster is chairman, and a large shareholder, in the Wigan Coal and Iron Company. The dispute arises out of the difference between weighing and measuring coal; the men demanding weight as most just than measure. He is also disliked for tyrannical strictness to the hands, and for calling upon the police to assist him in a recent trade difficulty. He is nevertheless a Liberal in politics, will go for the ballot, security to Trades Union Funds, and will give his support to Mr. Gladstone. If the miners could be persuaded to support him, it is thought that he would use his influence with the company to give or compromise the demands of the men. The miners have 3,000 votes*, but some of these are commanded by the Conservatives who have coal and iron works, but their influence is not great.'[32]

* It seems as if this is an overestimation of the miners' electoral strength. Probably he included those who lived in outlying villages, outside the constituency.

On 12 October, 1868, Howell wrote a letter, headed 'Private' to Mr Leigh Ellis Esq., Liberal Committee Rooms, Wigan:

> 'We are taking very important action in reference to Trades Societies, but we must deal with each Boro' separately and differently
>
> 'We are managing the unions in Blackburn, Bristol, Preston and a great many of the large towns at this moment and will, if you desire it, send down a man at once to do something in Wigan. At Sheffield our delegates have done what the local agents could not do, viz, unite the numerous trades into one committee for electoral purposes.
>
> 'The enclosed card will indicate the way in which we do our work. But in all instances we take independent action, so as not to commit the candidates in any way. They pay us by giving a donation to our special Fund, Mr. Morley, Treasurer.'

Two days later George Howell wrote to John Holmes, of Leeds, the colleague of Alexander MacDonald, to inquire who was the general secretary of the miners' union in Wigan. He was put in touch with Pickard, and a meeting of trade unionists was arranged to support the Liberal candidates.

This intervention undoubtedly played an important part in determining the outcome of the 1868 election. Recent research has shown that Conservativism was extremely strong in Wigan throughout the nineteenth century.[33] Foremost in Tory ranks was the Lindsay family, who provided Wigan with Members of Parliament 1820-31, 1845-57, 1859-66, 1874-80. Far from faltering in the period around 1868, Conservativism developed powerful roots among the working class. Professor Hanham says that in Lancashire local Conservative Working Men's Associations were quite often holding meetings that were attended by upwards of a thousand workers. The Wigan Operative Conservative Society was especially strong. It claimed to have an unbroken existence since the 1840s and its high degree of activity can be judged from the fact that it regularly held meetings three or four times a week.[34]

All this makes the Liberal victory of Lancaster even more remarkable. For many colliers, it must have been an agonizing decision to use their first-ever vote in a parliamentary election to vote for him. Their conduct, can be explained in terms of deference. While many of them felt themselves able to challenge Lancaster's decisions about colliery questions, as they were directly relevant to them and within their own experience, political questions were another matter. It was thought that these were issues affecting the whole of society, which were too great, too complex, to be understood by mere miners. The Wigan election of 1868 was a classical illustration of what two sociologists recently described as 'a voluntary abnegation of power by the working class in favour of an hereditary or a quasi-hereditary elite'.[35]

Writers who refer to deference usually treat it as an absolute category: either workers display deference or they do not. But an examination of Wigan politics 1868-74 shows this is not always the case. Sometimes there is an ambivalence, an interpenetration of opposites, that makes the position far more complex. For instance, the Wigan Trades Council was formed on 1 January, 1872, and one of its avowed aims was to secure the return of working men to Parliament.[36] A year later, however, at its first annual meeting John Lancaster presided.[37] William Pickard told the Trades Council, since Lancaster had become an M.P., he 'had shown himself friendly to the labouring classes'. Yet, within fourteen months, Pickard was to fight this 'friend of labour', trying to deprive him of his parliamentary seat. In his election campaign at Wigan, Pickard displayed both deference and defiance: deference to the social order since he went out of his way to show his support for its basic principles; defiance because, in a sense, he wanted to disrupt this social order—a miner out to displace a master.

In essentials, Pickard's programme did not differ from Lancaster and Wood. The *Wigan Observer*, a Liberal paper, said it agreed with Pickard's politics but could not support him because the Tories might win on a split vote.[37] Without a distinctive platform, Pickard discovered he had not the advantages of the official Liberal candidates but he had some disadvantages all of his own. Being a workman, many treated him with suspicion: Who would pay his salary if he became M.P.? Would he not be more the representative of a trade union than a constituency? The Tory *Wigan Examiner* accused him of being a 'red republican' and a friend of Charles Bradlaugh.[38] It compared Pickard's relatively menial social status with the exalted position in the social hierarchy of the two Conservative candidates, Lord Lindsay and Thomas Knowles, of Hatton Hall.

Since three of his four opponents were coalowners, it may seem strange that Pickard did not dwell on the problems of coal mining in the course of his campaign. But to have done so, in a way that specifically appealed to the colliers, would have entailed making utterances that were critical of the coalowners. As a believer in industrial harmony, Pickard regarded this as totally inadmissable. So, insofar as he made a case at all, he devoted himself to other issues in a way hardly liable to arouse support. For instance, he alienated Catholic voters, about 30 per cent of the total electorate, with his opinions on education. His election address declared, 'I look more to getting our children educated than any little distinction in the way they are taught reverence for their Maker.'

Although Pickard was confident, saying at his eve of poll meeting that eighty-five out of the 100 people he had canvassed had said 'yes', the result was far less encouraging:

Lord Lindsay (C) 2,493; T. Knowles (C) 2,401; J. Lancaster (L)* 1,883, W. Pickard 1,134; H. Woods (L)* 1,029. (A two-member constituency.)

Pickard's intervention helped to lose the constituency for the Liberals. At the same time, it showed there was no large, enthusiastic body of workers wanting to see members of their own order in the House of Commons. George Roby, an old Wigan Chartist, appears to have expressed the commonly held view of this campaign for labour representation: 'This new sect is mongrel in its composition, indefinite in its statements, and illogical in its pursuits.'[40]

At Merthyr, Halliday fared no better than Pickard. In a three-cornered contest he came bottom of the poll:

H. Richards (L) 7,606; R. Fothergill (L)* 6,908; T. Halliday (Lab.) 4,912 (A single-member constituency.)

Again, the result indicates the difficulty of translating industrial into political loyalty. Many of the constituents were colliers; the avowed aim was to unseat Fothergill, a coalowner with a tremendous hostility towards trade unionism; and yet the attempt, although successfull, still left Halliday apparently less popular politically than Fothergill.

There were three major innovations in the general election of 1874. The first, the one that has been universally recognized, is that the return of Burt and MacDonald to Parliament signified the beginning of a new era when working people could be represented in the House of Commons by members of their own class. The second, and less well-acknowledged, was the Amalgamated Association's political levy,

* Denotes previous sitting member.

as Mr Gwyn points out, 'set the pattern for all future trade union support of candidates'.[41] And, third, that the Amalgamated Association candidates stood as Labour candidates, entirely free from Liberal backing, and thus they can be said, in a sense, to anticipate the rise of independent working class political activity, which began with the Social Democratic Federation and I.L.P. But it is probable that the independent stand of Halliday and Pickard was largely fortuitous: the constituencies which they chose, both with a high percentage of miners, also happened to have Members of Parliament who were coal-owners as well. Hence there was not the same opportunity of reaching a detente with the local Liberal organization. It may have helped, in some people's eyes, to have given Halliday and Pickard an aura of extremism.

What was comforting to The Establishment, however, was the impeccable respectability and moderation of the two working men who had been elected as M.P.s. Benjamin Disraeli, the new Conservative prime minister, congratulated Burt and MacDonald on their success, and he continued:

> 'But all the sham working men candidates and trading demagogues who have pandered to the passions of the thoughtless for so many years have been rejected without exception by the sensible working classes of this country. I think it is one most satisfactory result of the general election. I know nothing more vile than an English Jacobin.'[42]

Equally satisfied with the result was Engels who, while he recognized Burt and MacDonald were extreme only in their moderation, recognized their victory represented precedent:

> 'the ice has been broken and two workers now have seats in the most fashionable debating club of Europe, among those who have declared themselves the first gentlemen of Europe ... the elections of 1874 have indisputably ushered in a new phase in English political development.'[43]

That new phase has been recognized by historians ever since.

CHAPTER NINE

Seven Lean Years (1874-1881)

The collapse of the A.A.M. inaugurated a period of decline. Disenchantment with trade unionism pervaded the entire industry—at pit, district and national levels. Colliers became more reluctant to pay their subscriptions; individual lodges, with dwindling financial resources, found it increasingly irksome to hand money over to district unions; and the national union discovered it was having growing difficulty with getting affiliated organizations to meet their obligations. Depression in the coal trade aggravated their predicament.[1] Their failure to arrest the decline in wages led to further enfeeblement. Miners expressed discontent with their organizations' inability to resist the coalowners' exactions by forming fresh organizations. Weak unions, unable to cope, became even more debilitated by splits, the proliferation of miniscule mining unions, which were quite incapable of meeting the challenge of the times. Demoralization and frustration was inevitable: membership fell as the number of unions increased. According to Thomas Ashton, by 1880 there were twenty district unions in Lancashire and Cheshire with a combined membership of only 4,000. One estimate puts the figure as low as 1,500.[2]

Circumstances, anyway, severely limited the men's bargaining power. Unemployment was rife. There were frequent 'knock days'—the practice of sending colliers home after they had completed three-quarters of a day's work, but only paying them half a day's pay. In June 1876, Lancashire coalowners demanded a 15 per cent wage reduction. Most miners offered merely token resistance, temporarily stopping work and asking for the matter to be referred to a board of arbitration. When it became obvious that owners would not brook such manoeuvres—R. Fletcher & Son of Outwood, quickly evicted colliers from their cottages—all organized resistance crumbled. In the ensuing confusion, men at the Earl of Bradford's collieries ceded from Farnworth and Kearsley district; Hindley left Wigan union; Worsley union dissolved, dividing its money among its few members.[3]

In contradistinction to the growing disunity among colliers, the coalowners were achieving a greater degree of unity than ever before. Hitherto their organization had been confined to south Lancashire and Cheshire, but the new body—the Lancashire Coal Trade Association, formed January 1877—embraced owners throughout both counties. Its main concern, initially, was the transport of coal, the failure of the railways to meet the requirements of the industry. The Association claimed that the accommodation points for shipment were bad; there was an inadequacy of engines and men; coal was in transit for upwards of a month; and an excessive time was taken to return empty wagons, which were often damaged. The Association's first declared aim was to take action against the railway companies, especially against the main culprit, the North-Western, for compensation.

Coalowners' attitudes had been hardened by frustration and litigation. The most celebrated case was that of a large Bolton coal proprietor, Thomas Gidlow, of Heaton, against the Lancashire and Yorkshire Railway Company. The point at issue was the Railway Company's right to refuse to handle coal unless it was put in ten-ton wagons in batches of fifteen wagons. After eighteen months of legal arguments, Gidlow won his case. The court ruled railway companies had no right to make favourites of certain customers to the detriment of the rest; their charges must be as authorized by Parliament and exactly the same as those paid by other owners. Victory had cost Gidlow £1,200 in legal expenses and the *Manchester Examiner* remarked, 'A man of more yielding tendencies, or possessing a shorter purse, might have had to retire from the courts before the final issue had decided.'[4] The newspaper contended that the railway companies held a quasi-monopoly situation, making it impossible for coalowners to take their business elsewhere, and hence it was vital they should not be permitted to abuse their immense power.

It is significant that the coalowners' association was preoccupied with the problems of transport; the infirmity of the colliers' unions meant they did not cause the masters much concern. When in October 1876, the unions did attempt to re-group, their effort was feeble and short-lived. The M.N.U. had prodded the local Lancashire unions into action by threatening not to re-affiliate them unless, like other parts of the country, they had a properly constituted county organization. But the newly-created Lancashire & Cheshire Federation never became effective. Its leading officials—John Wildman (chairman), Joe Dawson (secretary), George Riley (treasurer)—were unknown and inexperienced. When confronting its first serious challenge, it just disintegrated.

Nemesis came after a further wage battle. In May 1877, the coalowners decided to impose another 10 per cent reduction. In response, the men met at the Co-operative Hall, Manchester, to determine their attitude. A hundred and sixty delegates, representing thirty to forty thousand union and non-union miners, resolved to oppose the cut. They did, however, offer arbitration as an alternative, saying they would accept any decision reached. But the coalowners rejected this proposal. They even refused to meet the miners' deputation. Yet, despite this snub, the Federation's officials still journeyed to Liverpool, where the masters' were meeting, hoping to persuade them to relent. The owners remained unmoved.

When the lock-out was in its third week, cracks began to appear in the men's ranks. Haydock colliers returned to work and this started the drift back elsewhere. This was partly prompted by the inadequacy of funds: for every pound colliers received under the old A.A.M. arrangements, they now received only seven shillings and sixpence. Even making payment at the low rate, funds were quickly exhausted. On 14 July, the Lancashire & Cheshire Miners' Association admitted 'having arrived at a crisis'. Four days later, a Bolton delegate conference heard that the union fund was exhausted and that the response to appeals had been poor. By twenty-three votes to sixteen, delegates conceded defeat.

Paradoxically, a more defiant attitude was taken by the non-unionists. Since they received no strike pay, they had been entirely dependent on private generosity and money gained by doing odd jobs. Nevertheless, they remained resolute. A meeting of non-unionists in Wigan Square resolved to continue the struggle. One speaker talked on union leaders 'living in the lap of luxury'. Another, Robert Bankes, said 'He had been a worker in the pit and he knew what it would be were he to return to work—plenty of work and no money ... He would rather starve in jail for a month rather than give in.'[5] This display of determination was futile. Once the union had decided to surrender, no useful purpose could be served by further resistance. Within a week of the union's decision, everybody had returned to work.

The defeat not merely revealed the organizational weakness in Lancashire but also nationally. Had the M.N.U. provided financial assistance, then the plight of the local unions might have been less. Aggrieved colliers could point to Alexander MacDonald's pledge, given to the Manchester meeting that resolved what action to take. He had told the conference that the M.N.U. would give support if the owners refused arbitration.[6] Later, after the coalowners had rejected arbitration and even refused to discuss the matter, MacDonald changed his mind. He told the M.N.U. Durham conference (May-June 1877) no help should be given. The men were responsible for stopping work, without first gaining the authorization of the M.N.U., and consequently it would be unconstitutional to assist. In reply, Lancashire officials explained to conference that they had adopted an extremely conciliatory attitude. With documentary evidence, they showed that they were confronted with the alternatives of either accepting a unilaterally imposed 10 per cent cut or being denied the opportunity to work. Nevertheless, the Durham conference backed MacDonald. It was prepared to give sympathy to the Lancashire men, not money.[7]

For Thomas Halliday this decision was the final indignity. He had been the national secretary of the M.N.U. ever since the Leeds M.N.A.-A.A.M. fusion conference of 1875, but the Durham decision clearly revealed the failure of Halliday's efforts. He had been unable to push the new union towards a policy akin to that of the defunct Amalgamated Association. Indeed, the trend was in the opposite direction.

When he became M.N.U. secretary, Halliday had seemed to have a reasonable prospect of achieving this aim. The M.N.U. agreed to support strike action in instances where the owners refused arbitration. As with the A.A.M., the national executive would determine industrial policy. Speaking to South Staffordshire colliers in November 1875, Halliday assured them that arrangements for strike pay would be the same under the M.N.U. as under the old A.A.M., except they would now receive only 7s. 6d. from national funds whereas previously they had received ten shillings. However, he thought the difference would be made up from local funds.[8]

But events conspired to ruin Halliday's plans. The fact was that, even with money collected by the M.N.U., industrial disputes usually ended in failure. The owners introduced the wage cuts they had intended, as well as gaining the bonus of smashing, or crippling, the union in the coalfield involved. This is what had happened in Scotland in 1874, South Wales and Staffordshire in 1875, and Lancashire in 1877: unions lost most, if not all, their membership and funds in unsuccessful attempts to stop the decline in wages. Surely such struggles were counter-productive, argued the moderates; it was better to accept the inevitable. They wanted miners to acknowledge that unions were powerless to prevent market forces from dragging down wages rates. Instead of senseless strikes, the moderates wanted all efforts devoted to developing, co-operation and understanding with the coalowners.

Those who regarded strikes as suicidal pointed to the colossal cost and futility of the South Wales dispute of 1875. It had involved fifty-eight thousand miners and twelve thousand subsidiary workers for nearly twenty-two weeks. *The Merthyr Tydvil Express* estimated it cost:

Coalowners (sale of coal)	£2,100,000
Ironmasters	2,150,000
Railway companies	264,000
Royalties	150,000
Wagon hire	100,000
Dock dues	100,000
Maintenance of pits, horses, plant, etc.	250,000
Local trade	1,000,000
Workmen's wages	3,100,000
Relief Admin. by Local Guardians	25,000
	£9,239,000

Some of these estimates are open to question.

Lord Aberdare, in a letter to the *South Wales Daily News*, thought workmen had lost three million pounds; some sources placed the figure as high as five million pounds.[9] Whatever the precise amount, it did not alter the general picture. The conclusion to be drawn from the South Wales strike was that the union failed to prevent wage reductions and bankrupted itself in the process. For leaders like MacDonald and Burt, the lesson was obvious: no strikes.

Hence, when Halliday came to advocate supporting men in disputes, he encountered growing resistance. The M.N.U. executive looked on it with disfavour and, while never expunging the right to strike from its rules, it increasingly qualified such action with impossible conditions. The large county unions, led by moderates like Crawford and Burt, stated they were not prepared to see their carefully garnered funds squandered on strikes elsewhere. Since power inside the M.N.U. was vested in the various leaders at coalfield level—not in some central authority as had been the case with the A.A.M.—this effectively acted as a veto.

So Halliday, thwarted by the moderates and disgusted with the M.N.U.'s non-intervention in Lancashire, ceased to be the M.N.U.'s secretary and cut his links with the organization. However, when he turned to give his undivided attention to affairs in the North-West, he found them in a state of chaos. The Lancashire & Cheshire Miners' Association had collapsed insolvent; his own district, Farnworth and Kearsley, was greatly weakened and no longer prepared to employ him; and so, in September 1877, he terminated his connections with the industry.[10]

But Halliday still retained his interest in mining. He acquired a job that gave him the opportunity to journey about the country. Thus he maintained his contacts in the coalfields. Several times in the next three years he tried to create an alternative to the M.N.U. On 6 May, 1878, Halliday chaired a meeting at the Mechanics Institute, Manchester, attended by representatives from Lancashire, Staffordshire and Yorkshire.[11] The main purpose of the gathering was to consider Halliday's proposal for what he described as a type of national insurance plan. It was not intended to interfere with the rules of local unions or their autonomy. All that would be involved was the payment of sixpence a month per member—reduced to fourpence during the current depression—and local unions would be able to call on these funds in emergencies. But Halliday's plan proved to be abortive. Times were hard, and colliers were unwilling to increase their union contributions.

But, in any case, this kind of plan would have proved unworkable. Normal insurance is based on the principle of risk-sharing, the assumption that if one person's house burns down he can be compensated from the premiums received from other people whose houses are not. If everybody's were to burn down simultaneously, then this principle on which insurance companies operate would be untenable. Yet, this is exactly what was likely to happen with miners' unions: in depressions coalowners demand wage cuts, the men insist on resistance—and all the local unions would be calling for money at the same time. Such a scheme might only prove workable if there were a strong central authority to determine overall strategy, an idea that had become unpopular with the demise of the Amalgamated Association.

After further attempts to re-build a national union, Halliday appears to have become dispirited. In October 1880, he told the Hindley Miners' Association annual dinner that he did not intend to have anything to do with the latest scheme, proposed by the North Staffordshire colliers, to form a national federation. Notwithstanding this, he commended the idea and said he thought they were correct to make the struggle for better wages and conditions a central aim. At the same time, he showed his antipathy for MacDonald's M.N.U., which was not bestirring itself by responding to the mounting restiveness of colliers throughout the country: 'He had heard that there was to be a conference of the National Association in Manchester in January. That was well-nigh three months off, and that being so, it might be said, "Live horse, and you shall have grass."'[12]

This seems to have been the last meeting in Lancashire to be addressed by Thomas Halliday. Mr R. Page Arnot records that:

'Halliday in his later life went back to the scene of his industrial battles in Cardiff. The Rt. Hon. William Brace tells how, when he was a young miners' agent, he often talked with Halliday, who, if not forced to beg his bread like Tommy Hepburn, had to make a living by selling small stores like oil and ropes around the Welsh collieries.'[13]

Halliday had fallen victim to MacDonaldism, the belief that industrial pacifism—if necessary, capitulation—was better than industrial struggle. So far had MacDonald marched along that road by 1879, he even remonstrated with colliers at Barnsley who were locked-out:

'Mr. Frith, he said, had remarked that they must be organised to stop the tyrannical efforts of some of the employers. The men made the tyrannies; they carried the weapon with them and, after they had applied the weapon that flogged, they turned round and cried out when it was used on them.'[14]

MacDonald and his colleagues considered that the ideal of harmony could be attained by arbitration, a settling of differences through discussion. Thomas Burt, M.P., expressed this opinion in a letter to the *Wigan Observer*, when he said the Northumberland Miners' Association 'all believe that the burdens that press upon the coal trade could be removed by co-operation rather than contention. With this view, they supported every reasonable proposition of the coalowners, however unpopular this might make them with the men.'[15]

The unpopularity was sometimes very pronounced. In his letter, Burt admitted that four of the last five arbitration decisions had gone against the men. As a result, in June 1877, the men came out on strike against the advice of their leaders. Burt may have had this in mind when, at the M.N.U. Durham conference, held in the same month, he contrasted the position of leaders of early mining unions, such as Hepburn and Jude, with his own present-day position. After saying that the pioneers had fought against tremendous oppression, he nevertheless pointed to one compensation they possessed: 'they had not the hostility of the men they took the lead of.'[16]

But Burt and company were prepared to live with the curses from the men. However unpleasant, it was still bearable and preferable to the alternative course—to side wholeheartedly with the colliers, incur the wrath of the coalowners and quickly arrive at a disastrous end. Burt could claim, with some justice, to have attuned his tactics to the relationship of class forces. Its vindication, it could be argued, came when one compared the Northumberland colliers, affluent and possessing a strong union, with their counterparts in Lancashire, who were less well-paid and possessed no effective union. On the other hand, the Lancashire collier could maintain that Burt's policy was inapplicable in his county, where coalowners were much less inclined to settle differences round a table. Big firms, like Andrew Knowles & Son, whose conduct influenced the rest, were violently opposed to the existence of trade unions. They definitely were not prepared to accept arbitration, however much the miners may plead for it. In Lancashire industry generally, as Dr Hobsbawm has observed, 'new model' employers were hard to find.[17] This certainly holds true for mining, and obviously it could not fail to affect the colliers' attitudes.

Northumberland and Durham dominated the M.N.U., both because of their financial and numerical strength. They were the aristocrats of mining.[18] An aura of wealth surrounded industrial relations in the North-East, cushioning blows, providing greater manoeuvrability, so that employers and employed were more likely to resolve differences. But Burt and company were wrong in thinking that these procedures, which worked well in the North-East coalfields, constituted a universal panacea. When the M.N.U. strived to apply them, it ran into difficulties, losing the affiliation of organizations who thought nothing could be gained by belonging to a national union that eschewed all industrial disputes. As the M.N.U. afforded no protection, the Lancashire district unions, except Ashton-under-Lyne, left after the 1877 disaster.

While Lancashire miners did not believe as ardently as Burt and MacDonald in the efficacy of industrial pacifism, nevertheless they still wished to bridge the gulf between capital and labour. Far from courting confrontations with the owners, their aim was amity. It is in this light, as an attempt to reconcile the opposites, that their attitude to co-operative ventures must be viewed. They saw co-operation not so much in Owenite terms, as a means of creating a new society, but rather as a means of lessening tensions in existing society.

When the Miners' Hall was opened at 35 Millgate, Wigan, in May 1876, the Wigan Co-operative Society rented the downstairs half as a shop. It must have provided the Society with spacious premises for the upstairs part contained rooms for the miners' agent, secretary and checkweighman, as well as an assembly room with capacity for 180.[19]

But this was not the first contact of colliers with the Co-operative movement. In 1872, the Darwen Industrial Co-operative Society, wondering where to invest its surplus capital, decided to form the Darwen Mining Company Limited. It acquired a colliery at Whitebirk, near Blackburn, and confidently expected an annual return of forty per cent on capital invested.[20] But it ran into difficulties and only two years later did the drawing of coal begin. Nevertheless, the first 26 cwts. was 'carted by the Industrial Society through the streets of Darwen in triumph'.[21] This was the only moment of exhilaration in the colliery's dismal career. When, on 23 December, 1882, the Darwen Mining Company went into voluntary liquidation, the chairman declared, 'the Whitebirk colliery never had paid its way and never would'.[22]

A similar ill-starred venture occurred at Spring Vale colliery, Darwen. There the Company acquired a fourteen year lease, later extended to twenty-eight years, to mine eight acres for £8,000. Royalties worked out at the cripplingly high figure of two shillings and sixpence a ton. When the Darwen Mining Company went into liquidation, the Co-operative Wholesale Society took over the colliery and installed new management. This did not result in a change of fortune. In March 1883, the colliery was sold for £1,700, leaving a total loss of £9,000.

The Manchester Equitable Co-operative Society held a meeting on 2 August, 1873, to consider 'the advisability of taking united action for securing a needful supply of coal from existing collieries or by co-operative means'. Delegates attended from Bacup, Bury, Eccles, Failsworth, Heywood, Rochdale and Manchester. They decided to set up a seven-man committee to investigate. As a result, the United Coal Mining Society was formed. Its prospectus, issued in November 1873, stated it was 'a federation of co-operative societies and co-operative mills as well as of individual shareholders'. Shares cost five pounds each.

In June 1874, this Society bought the Bugle Horn colliery, Westhoughton. It paid about £27,000 according to Benjamin Jones—£32,000 according to the *Wigan Observer*—for sixty-four Cheshire acres that contained Wigan three, four and five foot coal.[23] The colliery was much bigger than the other collieries in Lancashire where experiments in co-operative production were tried. It possessed two modern shafts, and was expected to produce about one hundred thousand tons of coal a year. The sanguine Society thought it would make a twenty-five per cent profit.[24]

But in this, as in the other cases, the estimates were far too optimistic. Demand for coal was poor, and production was even poorer. In July and August 1875, output averaged three hundred tons a month. It rose in the period from October to December 1875 to between six hundred and eight hundred tons. By August 1876, it had averaged a weekly output of six hundred and twenty-seven tons, and peak production—one thousand two hundred tons weekly—was reached in June 1878. But this was nowhere near sufficient to attain breakeven point, a weekly output of three thousand one hundred tons. Bugle Horn closed and with it, on 20 April, 1878, the United Coal Mining Society went into liquidation.

The Society's collapse meant many suffered financial loss: with the Co-operative Wholesale Society it was £20,000 while Bolton Co-operative Society lost £12,000. But it was the countless small savers many of whom had invested their life-long savings, who felt it most keenly:

> Oh, Bugle Horn! Oh, Bugle Horn!
> Much you've from our poor backs shorn,
> Money from our pocket torn.
> May you, then, in future warn
> Us such wretched schemes to scorn.
>
> Hope and trust. Hope and trust.
> Lo, the chairman says we must.
> Work till we return to dust?
> Better so than slowly rust.[25]

The failure of all co-operative mining enterprises in Lancashire can be attributed to a lack of technical expertise on the part of those making the decisions. The purchase of ailing mines, the lack of competent managerial staff, the slipshod manner in which future demand and output were estimated—all contributed to creating the tragedies. But the overriding reason for failure was the general depression in trade. Well-run privately-owned collieries were being forced out of business after 1873; the co-operative mines were merely sharing the fate of many capitalist ventures during that period. Professor G. D. H. Cole remarked that in his book, '*A Century of Co-operation*' (pp. 162-3):

> 'It is easy to see in retrospect that the basis on which these ventures were started was fantastic, and that they could never have succeeded. It is easy to understand how mines bought on the assumption of the absurdly high profits of 1872 and 1873 became

sources of heavy loss when the boom was over ... Co-operators may have been silly, but they were no sillier than a great many reputedly hard-headed business men.'

Unlike some other mining unions, who actually ran their own collieries, the Lancashire district unions did not have sufficient funds to commit to such ventures.[26] They contented themselves with merely commending such schemes to their members. Anything, in the union leaders' view, was worthwhile that helped to bring capital and labour together, giving the men a share of the profits. With this aim in mind, Halliday not only advocated co-operative collieries, but also acquiring partnerships and buying shares in large companies.[27] The only obstacle was a practical one—lack of cash.

Another aspect of mining where union leaders strove to win the co-operation of capital was in taking some positive steps to mitigate the hardship caused by industrial accidents. The carnage in the collieries gave this a top priority. During the 1870s Lancashire, the seventh largest coal-producer, frequently had the highest accident figures. Miners usually filled two-fifths of the beds in Wigan infirmary.[28] What brought matters to a head was a particularly bad spate of disasters, which left local relief organizations unable to cope with the large number of widows and children.[29] As a result, mainly through the energetic advocacy of William Pickard, the Lancashire and Cheshire Miners' Permanent Relief Society was formed in 1872.

Why this issue is dealt with at this juncture is that, throughout the 'seventies', the Permanent Relief Society was like a time-bomb, gently ticking away, until miners' frustrations with it led to a tremendous explosion. The massive strike of 1881 (dealt with in the next chapter), can only be understood when considered in this context.

In his '*Miners Insurance Funds—their origin and extent*', George L. Campbell wrote:

> 'In 1869, 1870 and 1871, the South-West Lancashire coalfield was visited by a series of disasters, which has had no parallel in the history of mining ... The funds raised to meet the distress thus caused were wholly inadequate, for the very frequency of the appeals to the public caused them to pass by almost unnoticed.'[30]

In the immediate vicinity of Wigan, there were the following disasters:

		Deaths	Widows	Children
Hindley Green	26 Nov. 1868	62	22	55
Norley	21 Dec. 1868	7	7	26
Haydock	30 Dec. 1868	26	41	97
Haydock	21 July 1869	59		
Rainford	8-10 Jan. 1869	9	6	10
Park Lane	1 April 1869	31	15	27
Low Hall	15 Nov. 1869	28	13	27
Brynn Hall	19 Aug. 1870	20	9	19
Pearson & Knowles	6 Sep. 1871	70	35	86
Pearson & Knowles	20 Sep. 1871	5	4	7
		317	150	354 [31]

Obviously, the distress was extensive. A Wigan Relief Committee collected money from firms, churches and the wealthy. It hoped to gain sufficient funds to give each widow five shillings a week until she died or re-married, and also to give three shillings for every child until he reached the age of fourteen. But this provision was inadequate: clearly most of the bereaved would be compelled to beg for poor relief.

By the nature of the coal industry at that time, disasters occasioned extensive hardship, extending beyond a dead man's wife and children. For frequently an unmarried miner would be the bread winner of the family. Other members of the family who were unable to work down the pit because of injury, physical infirmity or age, became dependent on his wage packet. Hence his sudden demise involved pecuniary loss for people for whom the relief societies made no provisions, as this letter from 'The Aged Fathers of Haydock and Surrounding District' indicated:

> 'We, the aged fathers of Haydock and surrounding neighbourhood, who lost our sons and bread-winners in that horrible explosion at the Wood pit in last June, respectfully beg a small space in your paper. Some of us have lost one, two or three of our sons who would have held their place and helped to bring up the younger ones of our families, but we are now left destitute to mourn their sad fate. Some of us are in the lowest circumstances and compelled to apply to the parish for relief or otherwise become the inmates of the workhouse. We are not reckoned sufferers because the old fathers were not lost in the sad calamity.'[32]

William Pickard certainly expressed the indignation of the mining community when he talked of colliery disasters creating injustice, compelling people to 'become paupers unwillingly, and not through any fault on their part ... They ought not to be treated as paupers.'[33] He suggested that, as the problem was of such magnitude, a new organization should be formed, either on a county or national basis. Its funds should come from the colliers and coalowners, levied on each ton of coal produced. In his opinion, the proposed body should have three functions: to establish a permanent fund, build miners' orphanages and campaign for more stringent colliery rules.

Pickard's plan was received by miners with suspicion. When the union held a public meeting, few bothered to attend. The *Colliery Guardian* observed that a small committee room rather than a big hall would have been more appropriate.[34] The reason for this was that many colliers thought Pickard's scheme was likely to increase—not lessen—the number of accidents. At that time, most coalowners organized their own schemes. While realizing the benefits these provided were inadequate, the miners thought they at least had the redeeming feature of giving the coalowner a vested interest in safety. The more deaths and injuries at his colliery, the bigger the amount he would have to pay out. This incentive would vanish, however, if benefits came from a pooled fund. Then he would be able to work his colliery recklessly, consoled by the thought that others would pay for the deaths and injuries that ensued.

Miners also argued against Pickard that he started at the wrong end: prevention was better than cure. In their view, priority at that time should be given to a campaign for stricter mines regulations. To stop accidents from happening was more important than making provision for widows and children.

The miners' standpoint is understandable. At any moment, Lancashire looked likely to suffer a disaster of the same severity as that which occurred at the Oaks, Barnsley, where 334 lives had been lost. The Oaks explosion took place two years before that at Hindley Green and the *Sheffield Telegraph*'s editorial commented on the similarities between the two catastrophies. Re-assurances by the masters left the paper unmoved:

> 'Hindley colliery owners said their pit was in a less dangerous condition than the Barnsley one. It was not so great in the goaves; it had no places where the men who went packing had to "tumble out" for very life; and it had no blowers that roared, nor does it seem to have been the practice to let down the roofs. But it had floors that lifted, rifts that a man might put his hand in, places

where the oozing gas could regularly be heard "singing" ...'[35]

Commenting on the inquests after the Hindley and Oaks disasters, the *Sheffield Telegraph* concluded, 'The suspicion that the whole truth does not come out is irresistible.'

This appears to have been the case at most, if not all, the inquests in this period. Instead of being concerned to discover the cause of the disaster, so that a repetition might be avoided in the future, the actors seemed to be more interested in concealment. Colliery officials would endeavour to exonerate themselves; underlings would strive not to jeopardize promotion prospects; manufacturers would testify to the safety of their equipment.

Nevertheless, disquieting evidence occasionally did get heard. A large quantity of gunpowder, stored underground at Park Lane, Ashton-in-Makerfield, greatly increased the magnitude of the explosion in which thirty-one lives were lost. Henry Wright, a fireman at the pit, who should have reported this infringement of the regulations said he failed to do so because he feared dismissal.

Henry Wright: 'I have lodged complaints before.'

Peter Higson: 'And got the sack?'

Henry Wright: 'No, but it was reported I should be sacked.'

Wright then mentioned a Quaker collier, Matthew Heyes, who was victimized because he complained about the failure to observe safety regulations.[36]

A similar incident happened a few months later at the inquest into the Haydock explosion, where fifty-nine men died. A collier called Billinge said he had been dismissed for giving evidence, but the company claimed it was for stealing wood. Even so, his sacking had an inhibiting effect on subsequent witnesses, as proceedings show:

The Foreman: 'A rather serious circumstance has arisen. A juryman tells me that one of the witnesses called yesterday has stated he dare not tell the truth.'

The Coroner: 'What witness is that?'

A Juryman: 'Joseph Wright.'

Another Juryman: 'He told me the same.'[37]

Colliers wanted inquests to be genuine inquiries, unimpeded by the hidden hand of the employer. They also wanted stricter safety regulations, enforced with greater rigor than the prevailing ones. This was the way, in the opinion of many miners, the spate of Lancashire explosions could be ended. Towards that end, Pickard's plan seemed to them an irrelevance.

Coalowners, however, generally took a more sympathetic attitude to Pickard's scheme. Until then, most masters had run their own sickness and accident schemes. These had, for the employers, certain distinct advantages. They helped to lessen the mobility of labour since movement from one mine to another meant sacrificing one's benefits; they provided the owner with fresh liquidity in the form of workmen's contribution; they gave the employer, who took the key decisions, a further weapon he could use against employees.[38] Yet, despite these advantages, most employers were prepared to see their schemes wound up. The reason simply was that, while the schemes could cope with minor disasters, in the age of big colliery explosions their funds were insufficient to meet the demands that might be made on them at any time.

As a result, the Lancashire & Cheshire Permanent Relief Society was formed in 1872. Miners paid threepence a week. In return, they received eight shillings a week if off work through an accident. The widow of a man killed in the mine had a lump sum of £20, plus five shillings weekly and two and sixpence for each child under fourteen years of age. In the period 1873-8, 3·295 miners lost their lives per thousand employed and 18·75 miners were disabled on an average each year. But the greatest call on the Society's resources came from those colliers who suffered some temporarily incapacitating industrial injury. In 1879, C. G. Jackson, the chairman, reported: 'out of twenty-seven thousand members over five thousand had received benefit from the Society, showing that so many people had been more or less hurt during the year.'[39]

This meant that the average miner could reasonably expect to be involved in several accidents during the course of his working life. If unfortunate, he might be crippled or killed. As a result, his family would become paupers. This fear of the workhouse, a dread which was his constant companion, made miners exceedingly concerned about provisions for industrial injury. Even men who were apathetic on the question of trade unionism were militant on this issue and staunchly opposed to the Relief Society. Their first objection was that its control was firmly in the hands of the coalowners. At the top level, on the executive, twenty-three of its twenty-four members were coalowners and Maskell Pease, secretary of the masters' association was also secretary of this body. At colliery level, management was more diversified. 'At some collieries the underlooker and his chief firemen form both the committees and the officers.' In others, committees were constituted of 'creatures of the underlooker ... all nominated and elected by himself'. While in still further pits the Society was run by the colliery clerk.[40]

Whatever form it took, control was always firmly in the coalowners' hands. The miners resented this since they provided three-quarters of the finance; only a quarter came from the masters. It riled colliers that, whenever their interests clashed with those of the coalowners, the Relief Society was certain to side with the owners. Some miners even suggested that the Society connived with owners when infringements of safety regulations led to accidents.

Other criticisms included the high cost of administering the Society, the inefficiency of the administration and the poor service rendered in some instances. For all these reasons, many miners saw the solution to their problems in an industrial injury and sickness scheme run by the state.

But not everybody agreed with their view. William Pickard, as father of the scheme, was proud of the work it was doing. The situation, he argued, was far better now they had a Relief Society than before it existed. He therefore resented the criticism. Likewise did John Knowles, M.P. for Wigan. A collier who had risen to being a coalowner, Knowles' remarkable success can be attributed to the inordinate importance he attached to money. Believing that others shared his opinion, he told the House of Commons that if compensation were provided for injury, then some colliers would deliberately maim themselves in order to get the money.[41] He therefore opposed any bill that had such provisions.

Nevertheless, Parliament passed the Employers Liability Act of 1880. This constituted a direct threat to the coalowner-dominated Lancashire & Cheshire Permanent Relief Society. If workers considered the national scheme better, they could contract out of any private scheme. And this was precisely the move that Lancashire coalowners were determined to prevent their employees from doing.

Table 13 *Average daily earning of Lancashire miners*

	s.	d.
1811-20	4	3
1821-30	4	2
1831-40	4	0
1841-50	3	0
1851-60	3	8
1861-70	3	9
1871-75	6	10
1876-80	4	4

Source: A. L. Bowley, *Wages in the United Kingdom*, p. 118.

CHAPTER TEN

The Formation of the Federation

The 1881 dispute was the biggest and, to historians, least well-known strike in the Lancashire coalfield. Mr Page Arnot makes a passing reference, a mere sentence, while Thomas Ashton confines himself to outlining the main causes and stating that the dispute simmered till March: 'Pits opening and the men commencing work, and, after being at work a day or two, crowds of men from other districts in the county came along and surrounded the collieries and demanded the men be brought out of the collieries, which was done.'[1] Nothing in these two accounts conveys the importance of the battle or the ferocity with which it was fought. Yet, contemporary observers were in no doubt. Dickinson, the mines inspector, reported to the home secretary: 'A turnout of colliers, such as perhaps was never known in this district, took place in January and February.'[2] The *Manchester Guardian* was more explicit:

> 'The strike, which last week was practically brought to a close, has been the most serious dispute of its kind ever known in the coal trade of this district. As a general stoppage of work, there has been no strike in Lancashire which can at all be its equal, and fourteen years have elapsed since any previous dispute of considerable magnitude occurred in this coalfield ... the area of the recent struggle has embraced the whole of the Lancashire coalfield and, for two or three weeks, involved a complete stoppage of work on the part of at least 50,000 men and boys.'[3]

The *Manchester Guardian* calculated that it cost the coalowners two million tons of coal. At the same time, their competitors gained a foothold in the coal markets of Manchester and Liverpool.

Many issues intertwined in the dispute, causing chaotic confusion. Some men were fighting for one thing, others for something entirely different. So miners in one area, returning to work, might angrily collide with miners from neighbouring pits who regarded them as blacklegs. At any one moment it was impossible to say, with any exactitude, who was on strike and who was not.

Undoubtedly, however, the dispute was sparked off when most of the coalowners issued an ultimatum. They told their workers that they would either have to remain members of the Lancashire & Cheshire Permanent Relief Society or their employment would be ended on 1 January, 1881. This angered most colliers. Excited meetings were held throughout the coalfield, where speakers questioned why they should relinquish the rights to be gained under the new Employers' Liability Act.

Concurrently, the colliers had been considering submitting a fifteen per cent wage claim. Quite naturally, in the fresh situation, with the employers locking them out because they wanted the new legislation enforced, the colliers felt entitled to make their fifteen per cent an issue in the struggle. Then things spiralled. The owners retaliated by calling for the introduction of a sliding scale; some men widened the area of contention by demanding an eight-hour day while others shouted for payment of wages on a weekly basis.

Both sides suffered from lacking clearly thought-out aims and strategy.

In the workers' case, this position was aggravated by the diminutive size and influence of the district unions. These were too weak to provide a united voice, giving coherence to the babble of diverse demands. But the unions' inadequacies did not end with that: their membership was so meagre, their contact with colliers in many pits was limited, fragmentary, and consequently it was impossible for them to act as true representatives. By their very nature, miners' unions themselves produced a further obstacle to effective communication with the mass of the men working in the industry. This was because the unions tended to contain the aristocrats of mining, the better paid workers, the ones who did not share the anguish and anger of the most downtrodden.

As other writers have noted, checkweighmen had a profound effect on the development of mining trade unionism. Their job placed them in an admirable position, coming daily into contact with both master and men, to play a prominent part in any union lodge that was formed. Indeed, it seems almost inevitable the checkweighmen should become the main spokesmen of the miners. But, obviously, their dealings would largely be with the hewers who actually won the coal and whose wages were dependent on its weight. With them, there would be intimate, day-to-day association. But checkweighmen would have less to do with the rest of the labour force, men paid by the day, not by the ton. These others—daywagemen and datallers—generally received poorer wages and, since the checkweighman was the focus for trade union activities, they were less likely to hold a union membership card.

In normal circumstances, then, union affairs at the typical pit would tend to be dominated by the hewers paid by weight and the checkweighman. But, to some extent, this would be concealed by the large number of others who would be in the organization. However, in times of dwindling membership, the first to leave would be the daywagemen and datallers. As a result, only a skeleton organization consisting of the checkweighman and better-paid coal-face workers would remain.

The daywagemen and datallers, being less well off and regarding themselves as the most exploited, constituted an explosive mixture, likely to be more violent and less amenable to compromise. Almost certainly, it was this section that was responsible for much of the unofficial (i.e. not union authorized) activities during the strikes of 1868, 1874—and, again, in 1881.

In Lancashire, the non-unionists were frequently more militant than the union members. This was clearly shown in 1881. Leaders like William Pickard wished to avoid a struggle. Yet, they were faced with a decision, reached by representatives of Lancashire, Yorkshire, North Wales and the Midland miners, that in all areas where a sliding scale did not operate an application for a fifteen per cent increase would be made. William Pickard realized what the sequence of events would be if they tried to implement the conference decision in Lancashire: wage claim; offer rejected; a strike. So he sought to circumvent the resolution. Coal prices were starting to rise and, in many areas in Lancashire, the coalowners were anxious to introduce a sliding scale. Pickard calculated, that, if they negotiated a sliding scale agreement quickly, it would absolve them from taking action about the fifteen per cent. At the same time, however, under the new agreement, Lancashire colliers' wages would rise with prices. From William Pickard's standpoint, this stratagem might have a further advantage: in the convivial atmosphere created by higher wages, colliers might not grumble too much about being expected to remain in the Lancashire & Cheshire Permanent Miners' Relief Society.

William Pickard was an extremely powerful man in union affairs. He was miners' agent for Wigan, Pemberton, Standish, Aspull and Blackrod. These districts met to discuss Pickard's proposed package deal. When one or two delegates objected that his plan would mean miners losing the provisions of the Employers' Liability Act, Pickard replied, rather disingenuously, that his proposal would not involve contracting out of the Act since the Act itself specifically made the provision for alternative schemes to co-exist with it. Pickard gained delegates' consent, and, along with a deputation, went to negotiate with the employers in Liverpool.

But Pickard miscalculated. In making his deft manoeuvres, he failed to take into consideration the intensity of miners' hatred of the Lancashire & Cheshire Permanent Relief Society. Confronted with the coalowners' ultimatum, either to contract out of the Act or be locked-out, almost without exception chose the latter. The major portion of the Lancashire coalfield lay idle on 1 January, 1881.

This action astonished the coalowners. They had been led to believe that only a small minority, 900 out of 30,000 objected to existing arrangements being continued. They issued a statement, saying they 'heard with surprise and regret' the colliers' claim that they were coercing them. After friendly negotiations between masters' and men's representatives, it had been the suggestion of the men's representatives that workers should be expected to contract out.[4] The statement also said that weekly wage payments were impractical and, unless a sliding scale were accepted, there could be no increase in pay.

The employers' declaration made the miners extremely angry, not only towards the coalowners but also against the union leaders. One collier told a meeting at Leigh, 'They had a Judas Iscariot at the head, and that was William Pickard.'[5] At another meeting at Leigh, attended by 8,000 people 'the following resolution was carried amid great enthusiasm and hissing:

> That this meeting resolves and wishes the Press to convey to the public, that Mr. William Pickard, of Wigan, no longer represents the miners of Leigh, Bedford and Hindley Green district, and that he has no business with us hereafter.'[6]

The Leigh Chronicle explained that Pickard had not been employed by the Leigh district for the past five years: 'The so-called "dismissal" was simply an expression of the strong feeling on the part of the men.'

More disturbing for Pickard must have been that this feeling was shared by men from his own district. When, on 15 January, he rose to address a meeting in Wigan Market Square, 'groaning and hooting were freely indulged'. He denied he was 'a traitor and a man who was paid by the employers to do certain things'. But this did not satisfy the gathering. Thomas Oakes, a rank-and-file leader from Hindley, assailed him and proceedings ended with a fight.[7] This was the last time during the dispute that William Pickard attempted to address a public meeting.

On previous occasions when his authority had been challenged, William Pickard had always been able to count on leaders from other coalfields for vocal and moral support. But this time even his old friend, Alexander MacDonald turned on him, accusing him of being 'the catspaw of some designing party'.[8] Worse still, Ben Pickard, the Yorkshire leader, toured the Lancashire coalfield, speaking at Leigh, Tyldesley and Wigan to outline the benefits of the new Act.

The Miners' National Union conference, convened at Manchester on 10 January, heard that 'the strike, or whatever it may be called, had reached gigantic dimensions, and the stoppage of work is more general than has probably ever been known in the history of the trade of this country'.[9] While conference was careful not to support the strikers, it nevertheless deplored the employers' tactic (and, by implication, William Pickard's) of requiring men to contract out of the Employers' Liability Act. It was suggested that application of the Act should be made universal and compulsory.

William Pickard did not attend the M.N.U. Manchester conference. When challenged to explain why he was not prepared to defend his views at that gathering, he replied he had other business to do and, anyway, 'the conference was not held for the purpose of bringing about a combat between himself and Mr. Benjamin Pickard'.[10] This answer, as well as his conduct, left many miners very dissatisfied. His position was rapidly become untenable. At a Wigan delegate meeting in early February, 'Mr. Pickard was almost chucked out.'[11] He was removed from the position of chairman, and, in tears, agreed to tender his resignation as miners' agent.

Two months later, a *Wigan Observer* editorial reported:

> 'It will be remembered that during the progress of the recent strike, Mr. William Pickard intimated that owing to the course of action taken by the men, and more especially because of the personal abuse to which he was subjected, he should cease his connection with the union as its agent as soon as the district had settled down. This course he has now taken, and he has become the commercial representative of the executors of the late Mr. Diggle at Westleigh. Of the reputation Mr. Pickard has gained as a man of business, there can be no better evidence than he was selected for the appointment from 150 applicants.'[12]

Besides William Pickard, the union organization shared the discredit in the eyes of the majority of miners. They were frequently regarded as 'the self-chosen committee of delegates, who meet from time to time in the Miners' Hall without our knowledge or consent'.[13] Such a committee could not hope to control affairs. Insofar as there was any control exercised, it came from a heterogeneous group, spontaneously thrown up by events. Men like Noah Stringfellow, a dataller at Gidlow and Swinley; J. M. Forster, a former pit-boy who subsequently became a novelist;[14] John Cross, of St Helens, who had forty years of mining experience; Thomas Oakes, of Hindley, who had been collier and drawer for thirty-five years; and Thomas Parr, of Tyldesley, who, 'though he had been called a boy, he had during the last three weeks been leading the men in his district'.[15]

These, and many others, addressed large strike meetings, which were the principal means of boosting morale. They were held on average twice a week in each township. While the grievances expressed varied from meeting to meeting, certain themes were common to them all:

1 Opposition to the Lancashire & Cheshire Miners' Permanent Relief Society. 'They (the men) derived no benefit from it, and they had nothing to do with the management of it.' (N. Stringfellow, Wigan Market Square, 15 January.)
2 Opposition to the sliding scale. 'The masters had got them to the workhouse door, and another slide would push them in.' (W. Chantrey, Amberswood Common, 26 January.)
3 A determination to fight to the last. 'He had a wife and six children at home, and rather than go to work again without winning the battle, he would stand in the street and starve.' (John Houghton, Newton Common, 31 January.)
4 A willingness to use violence. 'They would have all men out either by fair means or physical force.' (Peter Cain, Leigh, 4 February.) 'The man who would "knobstick" was not fit to live. He ought to be sent into eternity.' (J. Tinker, Mosley Common, 27 January.)
5 A determination to win a wage increase. 'The miners earned the least wages and had to face the greatest dangers. They were being robbed by their employers in ten thousand forms, and there was scarcely a single point affecting their interests to which he could turn in which they were not most shamefully abused and robbed by their master.' (T. Storey, Amberswood Common, 26 January.)

6 Criticism on the miners' union 'The money paid the Society was foolishly and extravagantly squandered ... No attempt was made to enrol non-members ... I am also told that the agent of the union is a shareholder of colliery property.' (N. Stringfellow, Wigan Market Square, 31 January.)

Their concern over lack of democracy is clearly illustrated by the Wigan Market Square meeting of 2 February, attended by between four and five thousand. George Leicester, the chairman, complained:

> 'There has been several delegate meetings called together, but he thought that those who had attended had not represented the true heart of the men. (A voice: "They have not".) They had been self-elected delegates, pick-sharpeners, checkweighmen and shop-keepers. (Hear, hear and laughter.)'

Then the next speaker, Noah Stringfellow, grumbled because the time of delegate meetings and what transpired at them was kept secret:

> 'Had their delegates ever given notice at what time they would meet them, and given them the choice of selecting a delegate of their own choosing, and had they ever asked them what their opinion was that they might represent them in honesty at their next delegate meeting? ("No, No", "They never have", "Never".) Had they after their return held meetings and informed them regarding particulars of the business transacted? ("No, No".) Very well, then, he moved a vote of censure upon the delegates for not complying with the duties of representative.'[16]

On 15 January, when delegates were deliberating over the most recent letter from the employers, a crowd of men burst into the Miners' Hall and effectively ended the discussion. There was an attempt to keep all subsequent delegates' meetings a secret, perhaps to stop a repetition of the disturbance.

It is difficult, in view of William Pickard's silence throughout almost the entire struggle, to give a comprehensive account of the union's official view. Attempts by executive members to explain their position were usually drowned by shouts of 'Purr him off the cart' and 'Thou'rt sweetened by the masters'.

The only extended statement of the union's view came from Sam

Woods, who was Pickard's right-hand man. He deplored the strike, saying 'we are living in an age of retrogression'. He pointed out that the legally appointed delegate meetings of 12 and 17 January had both favoured accepting the coalowners' offer of a sliding scale. The deputation that went to meet the owners in Liverpool on 14 January favoured accepting their terms:

> 'I defy any man to contradict when I say that if you had listened to me and the other parts of the deputation when we returned from Liverpool and advised you to accept the masters' terms, viz., To resume work on the Employers' Liability Act which they promised, and also that they would meet the men's duly authorised representatives as soon as they could get ready to draw up a fair and reasonable sliding scale, and, if it warranted an advance they would concede it and consider the question of weekly pays, which I believe, they will give the men the advantage of—if, I say, you had acted on the advice given, your position today would be a hundred-fold better than it is.'[17]

Sam Woods claimed that the real representatives of the men were the checkweighmen, 'selected for their intelligent and sober habits'. He argued that 'a dataller, who has nothing to do with the collier, has not the least shadow of a claim to go to the different mass meetings and spout'.

One feature that characterized the dispute was the widespread use of violence, as a means of persuasion. Men, their faces covered with scarves and smeared with coal-dust, roamed around attacking anybody who had the effrontery to return to work. Often music, played by bands in the middle of the night, would provide war-cry; the men would assemble; and then would march in the dark to a nearby colliery to make a dawn assault.

The violence, far from being just wanton destruction, had certain definite objectives. Wagons, office accounts and railway-lines were destroyed to stop the removal of coal on the pitbank; action against strikebreakers was to stop further coal being hewed. It was argued that, the more coal that came on to the market, the more protracted would the dispute be.

Men were naturally angry against those who, in their opinion, prolonged the struggle. For Lancashire experienced an extremely cold spell in January 1881, with as many as twelve degrees of frost, and families suffered severe deprivations. Sometimes furniture was burnt to provide a little heat. Even so, the majority of people, most of the time, had no fuel. Children went to school cold and having had nothing to eat. Reporting on the situation in Wigan, the *Manchester Guardian* stated many had resorted to begging:

> 'The little children who ask for alms tell of mothers and children at home starving; the mothers tell of their children, having "nothing whatever in the house to eat". Such is the present condition of affairs that persons who have lived in Wigan say they have never seen anything approaching the present distress. Men who have been connected with every strike and lock out for the past 30 or 40 years confess that they never beheld such widespread suffering.'[18]

In such circumstances, the temptation to return to work must have been difficult to resist, particularly when some coalowners offered to pay wage increases and permit men to work under the Employers' Liability Act. There were even instances of concerns, like the New Lester colliery, who stated they were prepared to give greater increases than the men had asked for. Moderate union leaders, such as R. Isherwood, of Tyldesley, argued that the men should accept. He thought it was wrong to penalize the small coalowners, who were fair and reasonable, because of the obstinacy of big firms like Andrew Knowles.[19] But the militants pointed to the lessons of previous struggles: when colliers at the small mines returned to work, it lessened the solidarity and broke the workers' united front. Also, since the large coalowners set the pattern for the rest the small coalowners often withdrew the concessions once the struggle was over. The result was everybody lost through a partial resumption of work.

The first acts of violence occurred during the first week of the strike. Men from Tyldesley, Leigh and Atherton, where a combined total of seven thousand men were out, marched to Boothstown. Singing 'We won't give in to th' masters upon their cheeky terms' to the tune of 'We won't go home until morning', they brought out colliers still at work, and proceedings ended with a riot.[20] Then the Wigan men marched to Over Hulton to stop the men there from working.

In the third week disorder became even more widespread. Haydock men threatened St Helens men with violence if they continued to work out their notice. As a result, all collieries in St Helens closed. Ten thousand strikers met in the town centre, the biggest colliers' meeting, according to John Cross, ever to be held in St Helens. Meanwhile, Leigh miners marched to Abram Coal Company, where thousands of tons of coal awaited dispatch. All the men working on the pitbrow 'laid down their tools and scampered off at full speed to their homes'. The Leigh contingent, armed with crowbars, pit props, bludgeons and sticks, quickly disposed of the police, who sought to bar their entry. Then, says the *Wigan Observer*, they swarmed on to the pitbrow:

> 'The police retreated out of sight into the locomotive engine house. A scene of tumult, which almost baffles description, then took place. Every tub of coal on the brow was turned upside down, every window in the huts and offices were smashed, and the mouth of the pit was surrounded by excited men, who declared they meant to go down and fetch up those who were supposed to be at work. The managers were bewildered. They assured the men that nobody was in the pit.'[21]

Seven men stripped off their clothes, went down to investigate, and returned fifteen minutes later satisfied.

A similar thing happened at Bickershaw Coal Company. A crowd of strikers said 'they would wreck the concern' unless men still at work were brought out of the colliery. The police advised the management to comply. Sixteen knobsticks then came out, and three strikers searched the underground workings to see that there were no more.[22]

At the Downhall Green pit, of Richard Evans, a dawn attack caught the Company unprepared. They had to agree to bring out the colliers still working. But, as this was being done, the police arrived, drew their truncheons, and a fight ensued in which three policemen were injured.[23]

At Chapel House Colliery, Skelmersdale, the police arrived on 21 January and 'found a mob of seven hundred literally in possession'. The strikers threatened to cut the winding rope if the strikebreakers were not brought up. The police advised the management to refuse. A running battle developed for over a mile, followed by two days of rioting in which Skelmersdale strikers were assisted by men from Rainford and Pemberton.[24]

Wigan colliers encountered little opposition when they journeyed to Blainscough, Coppull, where they stopped strikebreakers from working and did considerable damage to colliery property. But they did meet stiff resistance when, on 25 January, they marched eight abreast, in a column a mile long, with the intention of reaching Mosley Common, where some men were still working. *En route*, they called at collieries, assuring themselves all labour had stopped. At the Combermere pits of Tyldesley Coal and Cannel Company, the *Leigh*

Chronicle said 'the front ranks scaled the barriers like men at home at their work and knew every nook and cranny'. No knobsticks were found, and hence no damage was done. However, at the Rams' pit, owned by Wharton Coal Company, there was fierce fighting between police and strikers, which only ended when the manager brought out all the knobsticks. The *Leigh Chronicle* said that nearly every policeman was hurt in the engagement while the *Bolton Evening News* reported police injured at twenty.[25] Most had been hit by pieces of coal. In a letter to the *Leigh Chronicle*, a collier explained:

> 'Our lads could not get up stones or pieces of wood or anything to fight with. They were all frozen, or the handful of police there would have been overpowered, and possibly some of them killed, not on purpose (for the colliers have no ill-feeling against them), but in the excitement.'

In the battle an 18-year-old collier, Samuel Findley, of Chowbent, was killed while fleeing from the police.

The following day, 26 January, a mass meeting was held at Amberswood Common, situated between Wigan and Hindley, the *Wigan Observer* said it was the biggest gathering of its kind ever held in the locality. Thousands of men marched for four hours, through the fog and snow, from St Helens. They interrupted the journey at the Gravel Hole, Middle Place and Sawmill pits, belonging to Ince Hall Coal and Cannel Company, to destroy 50 yards of rail on the wagon way, an air-machine, windows and furniture.[26] At the meeting itself, plain clothes policemen were present. Almost certainly, better dressed and better fed than colliers, they must have been easy to pick out, and many miners, still smarting from injuries sustained at Wharton Hall the previous day, seized the opportunity to square accounts.

The authorities endeavoured to curb the colliers' lawless activities. The Lancashire County Constabularly had a force of 1,257 men, of whom seven hundred were involved in the dispute-affected districts. But these were inadequate, as General Cambron pointed out in a letter to the Lancashire annual sessions, to cope with the strikers, who were estimated to number fifty-six thousand.[27] As a result, aid from the military was deemed necessary.

The strikers do not appear to have been unduly impressed, however. The arrival of the Hussars at Leigh was greeted by 'We're not going to be beaten in by cabbage soldiers'. The proclamation forbidding processions and begging were simply disregarded. And, while strikers thought the impressive show of military might was to frighten them back to work, leaders like Peter Cain still uttered the customary threats despite the presence of troops: 'The men working at the pits had been asked to come out quietly. They ought to come out, and, if they did not come quietly, they must use physical force.'[28]

The most serious trouble occurred at Leigh on 4 February, when ten thousand men, assembled in the New Market Square. The *Leigh Chronicle* reported that 'the dense mob amused themselves by cheering vociferously and waving their sticks, pick-handles and other weapons', a stratagem which at least had the advantage of persuading some shopkeepers to give money and goods.[29] The crowd then marched to Chowbent and Tyldesley while the authorities tried to assemble sufficient forces to prevent Atherton being attacked. When the march reached Howe Bridge, between Leigh and Atherton, they had become 'quiet, listless and apparently harmless'. But it was there that the authorities decided to disperse them with a pincer movement, the police charging down Colliery Road as the Hussars galloped through the fields. 'Many in order to get out of the way, took refuge in the closets and outhouses behind the neighbouring cottages, from which they were hauled by the police and hammered.' Other colliers, however, retaliated by making repeated attacks on Lovers' Lane pit. After being beaten off by the police, they re-formed and charged again.

At the request of Captain Burn and Lieutenant Van-Votten Pole, of the Hussars, the magistrates agreed to read the Riot Act. But as Jabez Johnson was reading it, the crowd shouted, 'What the——do we care?' The Hussars then drew their swords and charged in all directions. The police also charged, and 'drew blood at every blow from those within reach'. It appears that they were more brutal than the military. Young boys were struck down, as were those who sought refuge in houses.

Nevertheless, the authorities did not gain victory instantaneously. At three o'clock, thirty knobsticks left Crombrooke pit, and were freely stoned. It was only by six o'clock that the streets were finally cleared of strikers. Yet, there is no doubt that the authorities succeeded. They had not only broken up the miners' march, but also stopped it from achieving its main objective, namely to stop men working down pits in the Atherton area. The following day the numbers at work were: Howe Bridge 100, Lovers' Lane 47, Chanters 88, Mr Ramsden's Shakerley 29, Gibfield some.

To commemorate the Hussars' victory, R. Greenhough, jun., brassfounder, struck a special medal with the inscription 'The Battle of Howe Bridge'.[30] When, on 1 March, the Hussars ultimately left Leigh, local businessmen presented Sgt. Major Frillcher, of the

Hussars, with a silver-mounted, hand-painted riding whip. The commanding officer replied: 'He carried with him the remembrance of pleasant times, and he jocularly remarked that he would ever entertain a lively recollection of the Battle of Howe Bridge.'[31]

Some Members of Parliament adopted a less composed approach. T. P. O'Connor, M.P. for Galway, drew the Home Secretary's attention to the serious disturbance 'during which there were twenty thousand cases of aggravated assault, twenty thousand riots, twenty thousand of resisting the local authority, twenty thousand cases of intimidation, twenty thousand cases of injury to property—and thousands of other offences, making a total of one hundred and forty thousand outrages in the course of a single day, and whether in the circumstances he would advise the introduction of a peace preservation bill, making every man and woman in Lancashire liable to 18 months' imprisonment at the discretion of the Duchy of Lancaster'. After early troubles, M.P.s had pressed the Home Secretary to adopt a tougher line. Daly Clarke, M.P. for Cork, argued that the Tyldesley riot had shown that existing means of coping with trouble were inadequate, and C. S. Parnell suggested the police should be armed with buckshot.[32] The home secretary replied that the action of the authorities, using existing laws and backed by public opinion, would prove to be sufficient.

The last serious disturbance happened when the dispute was in its seventh and final week. This took place at Ashton-in-Makerfield, New Springs and Hindley. At Hindley, the strikers were parading peaceably through the streets, to dispel rumours that they had returned to work, when the police approached. Then men stood on each side of the road to let the police pass, reported the *Wigan Observer*.[33] But, when they were in the middle of the men, they charged with their truncheons and cutlasses. 'Instead of our being the lawbreakers, it would be nearer to the truth to say the police were the rioters,' commented one striker.

The Hindley violence shows how, as a result of tensions built up during the dispute, attitudes can be transformed from sympathy to hatred. Only three weeks previously, the *Wigan Observer* had remarked:

> 'It is gratifying to learn that the populace of Hindley, and the military and the police pull well together, the greatest good feeling existing, and not a single disrespectful word has been said to those in uniform ... Every day the policemen make contributions among the ranks in aid of women and children.'[34]

'An Awkward Passage' was also published in *The Graphic*.

A further illustration from *The Graphic*, gives some idea of the laborious work it was before mechanization to sink a shaft.

In the midst of the seven-week struggle—indeed, arising out of it—came attempts to create a unified organization. A Leigh meeting, in early February, resolved to form a new union 'for the purpose of raising a fund to defend themselves in case of accident, and to take their claim to a superior court if need be'. Its leader, William Cambridge, stated it would be open to all men who worked in the pits to join. Compared to previous unions, it would have the advantage of regularly presenting balance sheets to its members and conducting its business without secrecy. But Leigh, a place where distress was felt with extreme acuteness, really did not possess sufficient financial resources to give the new union a sound basis. It is not surprising the *Bolton Evening News* of 1 March stated, 'The greatest difficulty of all is the collection of subscriptions.'

It was from areas that were less ravaged by the dispute that the impetus came that led to the formation of the Lancashire and Cheshire Miners' Federation. Like a bowl of porridge, it was the edges that cooled first. The places least likely to receive visitations from irate miners, anxious to see a cessation of work, could devote their energies not only to arriving at an agreement with the coalowners but also establishing a firm foundation for union organization elsewhere. Ashton-under-Lyne and Oldham miners were on strike for a mere three weeks, returning to work on the basis of a 12½ per cent wage increase and the right to benefit from the provisions of the Employers' Liability Act.

Thomas Ashton, who had been general secretary of the Ashton-under-Lyne district since 1879, was the prime mover in the formation of the Federation.[35] He convened a meeting on 18 January, which decided to hold a further meeting that they hoped would be more representative. This second gathering, held on 15 February, resolved 'that Lancashire be organised as a Federation of districts. Each district to contribute to a central fund for certain objects to be decided upon.'

The inaugural conference actually took place at Manchester Old Town Hall on 19 April, 1881. Nine district unions were represented: Ashton-under-Lyne, Oldham, Skelmersdale, Worsley and Little Hulton, Atherton, St Helens, Hindley, Radcliffe, Bolton. Two other districts—Platt Bridge and Over Hulton—joined a month later. Even so, six district unions remained outside the Federation. They were Wigan, Pemberton, Park Lane, Haydock, Little Lever and Blackrod. Sensing that Wigan, the largest of these not affiliated, was the key to recruiting the rest, it was decided to hold the next conference at the Miners' Hall, Wigan, on 5 July, 1881.

Supporters of the new Federation pointed to the lesson to be learnt from recent experiences:

> 'The proposal to establish such a scheme has resulted from the struggle which has recently taken place between colliery proprietors and their employees with respect to the Employers' Liability Act, the want of a miners' association which would secure unity of action to those on strike or locked out having been greatly felt.'[36]

They pointed to the outcome of the struggle, the great diversity of settlements. Some miners had only gained a 5 per cent increase, others had won 15 per cent; some had achieved the right to work under the Act, others were still compelled to belong to the Lancashire & Cheshire Miners' Permanent Relief Society. The wide variety of agreements reached was sure to spell trouble for the future, it was argued, unless all could be brought under the unifying aegis of the new Federation.

The idea of Federation met with resistance. In some quarters, the desire to retain complete local independence remained strong. But others failed to join because of personal poverty. Another obstacle was the existence, albeit in attenuated form, of a rival union based on Leigh. The two bodies differed fundamentally in outlook: one trying to recruit membership on a restricted basis, the other trying to become a general union; one was moderate while the other was militant. The leaders engaged in mutual recriminations. Robert Isherwood, the Federation's first treasurer, was the full-time miners' agent for the Tyldesley, Atherton and Leigh area. So, in his day-to-day activities, he found himself clashing with those who were trying to set-up the rival organization. The strong dislike for Isherwood, on the part of the Leigh-based union, did not extend to the Federation's secretary, Thomas Ashton. He was a taciturn man, an efficient secretary and administrator, not given to making utterances that enraged people.[37] Also, his centre of operation—Ashton-under-Lyne—kept him away from daily contact with the rival union.

Strong objection, however, was taken to Thomas Aspinwall, who was elected president at the Federation's inaugural meeting. 'He was a shopkeeper and checkweighman,' his critics argued. 'The interests of such men were not identical with those of the colliers.'[38] The same point, namely, that he was a checkweighman and shopkeeper, could be levelled at Sam Woods, the next man to be Federation president. Both Aspinwall and Woods were ultra-respectable. They belonged to the chapel-going fraternity as well as being total abstainers. Woods even accused a collier of being a socialist because he wanted Wigan Public Library open on Sundays![39]

It was argued, by their opponents, that men like Aspinwall and Woods would not be regarded as truly representative of colliers as a whole. In fact, this was a great asset, the secret to their success. They were able to assuage many of the coalowners' fears, and arrive at agreement with them. By contrast, a militant leader like Noah Stringfellow would find coalowners not prepared to negotiate with him; far from wanting to see him representing the men, the masters would want to see him blacklisted from all mining jobs.

A revealing insight into the qualities looked for in a miners' leader of that period comes from a contribution to the discussion on who should fill the vacancy caused by William Pickard's resignation:

> 'In the first place, he must appear respectable so that when sent to the masters, it will not be a disgrace for them to speak to him. He should also be a sober man, so that he will not bring evil on the cause he propogates. Honesty is also necessary in this case, so that he may be entrusted with the money he may have to convey from place to place. He must be able to command the pen, as there are so many letters to be answered, and able to deal with figures, so that he can examine the Society's accounts for himself. It will be well for the miners if they can get a man who will stand between the master and the men and to give reason with them, that is to get both sides to give way a little ... The man appointed will have to be a public speaker.'[40]

Sam Woods, who gained the appointment as miners' agent for Wigan, possessed these qualities. So did the Federation's leadership generally. Indeed, the fact that they constituted a solid, dependable core played a large part in the Federation's survival and subsequent growth.

CHAPTER ELEVEN

The Search for Agreement (1881-1890)

The 1881 dispute was a chastening experience. Neither employers nor workers wished to see a repetition. Besides the suffering, sacrifices and financial losses, both sides had a disquieting feeling of not mastering events. The dispute had unleashed forces beyond their control. Therefore, to see that it did not happen again, industrial harmony had to be preserved.

Each side had their own special reasons for desiring a period of tranquility.

The Lancashire and Cheshire colliers had created a very rickety structure, a Miners' Federation, that was liable to collapse at the first gust of conflict. Its membership was unimpressive and many small unions, wishing to retain their precious independence, remained aloof. So the newly-formed Federation was dogged by the same impediment as previous attempts—the North-West miners' intense love for localism. At the same time, whatever progress was made in Lancashire and Cheshire was placed in jeopardy by disunity nationally, the failure of colliers throughout the country to evolve a policy to be implemented in the whole of Britain. Hence, Lancashire colliers were haunted by the fear that in some future strike their efforts would be imperilled by the influx of coal from other areas, just as they had been in 1881.

What had made immediate matters worse in 1881 for the Lancashire miners was that, after the dispute was over, the men were impoverished and the union funds exhausted. One veteran stalwart observed, 'I am sorry to find that the miners in Wigan and the surrounding districts are in a worse condition, as far as organisation goes, than I can remember them to have been in. It seems as if there was a sort of civil war going on among them ... It may be said of the Wigan district that it has had a name and a power in the past, but now, alas, it may be compared to a ship with a torn sail and shattered deck, having no captain, no first, second or third mates.'[1]

These remarks are not exaggerated. The 1881 strike had torn the union asunder, generating intense hatreds and mutual recriminations.

The position of Pickard, as miners' agent, had become untenable. Yet, when the union came to fill the vacancy, it only led to fresh dissension. A victimized miner wrote to the local paper. He pointed out that the two candidates from Hindley and Ince 'during the late strike ascended platforms and protested, despite probable consequences, against the demands of the masters' while the candidate of the moderates had been conspicuously absent: 'Personally, I should be elated to know what Mr. Winstanley was doing during that critical period.'[2] To which Winstanley later replied, 'He says I thought silence the best policy; what I have to say is that it was not safe to be anything other than silent.'[3]

Because of his unpopularity, Winstanley was prevailed upon to withdraw. The candidate for the moderates then became Sam Woods, despite the fact that he lived outside the district and consequently had his nomination queried for this reason by the three other candidates. But, with the chaotic state of the organization, it was very difficult to apply the rules. Who were members? What other districts continued to exist? What were the boundaries of them if they did exist? These were questions open to a variety of answers.

Sam Woods secured the post. In accepting the agency, he made a speech that, in an intriguing way, showed a distinction between him and his predecessor. Whereas William Pickard had told a Liberal meeting at the 1880 general election that fighting 'the naked savages of Zululand' was 'rather degrading besides landing us in great financial difficulties', Sam Woods likened union-building to empire-building: 'By combination bright and precious laurels have been won by British soldiery when they had left the home on their native shores and gone and invaded foreign territory.'[4]

There was definitely a need for something to unify the warring territorial groups. Skelmersdale had left Wigan, claiming it had been tricked. Park Lane thought it could do better on its own. Hindley considered that all it required was its own checkweigh and labour fund. Blackrod and Standish had, at best, an ambiguous attitude towards Wigan. Other components of the district union may well have ceased to exist.

The disarray in central Lancashire seems reasonably typical of the county as a whole. Lack of leadership and organization prevailed throughout. In such circumstances, it would have been a foolish and reckless act to seek out confrontations with the coalowners.

Likewise, with trade improving, the masters wanted to avoid conflicts that would interrupt production. In order to preserve industrial peace, they tried to achieve a *modus vivendi* with the men. While keeping their control over the Lancashire & Cheshire Miners' Permanent Relief Society, the masters endeavoured to make it more acceptable to the men. The management committee was enlarged to increase colliers' representation; a commission of two and a half per cent was given to local agencies, thus giving some men a vested interest in existing arrangements; the Society openly acknowledged a need to change with the times: 'The passing of the Employers' Liability Act produced a state of affairs which appeared very unfavourable to the progress of a movement depending largely for its success upon the maintenance of goodwill between employers and employed.'[5]

Conciliation also became the key-note of the coalowners' approach generally. They conceded ten per cent pay increases in the autumn of 1881 and again in 1882. Instead of fighting the entire mining population, the employers confined themselves to limited engagements, whittling away allowances or taking probing action against small groups of workers. The Crawford collieries made an attempt to introduce Billy Fairplay, a South Wales system for weighing coal. This was unpopular among the miners. A wider mesh was used, a greater amount of coal fell through, and hence colliers paid by weight received less money. A four-month strike was required to stop its introduction.

Another large coalowner, R. Evans, decided to have a trial of strength with the union. He gave notice of a ten per cent wage cut at three of his thirteen pits. The Federation responded by sending a deputation to the general manager to ask for the matter either to be referred to arbitration or for the cuts to be withdrawn. But he replied that he could not entertain the idea—indeed, would rather lose his right-hand than give way. The Federation then resolved to withdraw labour at all thirteen pits, levying the entire membership to provide strike pay. Two months later, work resumed with no cut whatsoever. This was a victory for the Federation. Thomas Ashton claimed that the prolonged payment of strike benefit was 'a feat which had never before been accomplished in Lancashire'.[6] While this is historically untrue, Ashton was nevertheless correct to claim it as a sign of strength. Many colliers were certainly impressed and sufficiently encouraged to join.

Not only did the Lancashire membership increase, but the North Staffordshire colliers applied to join the Federation. At first glance, this may appear rather strange. It must be remembered, however, that the main market for Staffordshire coal was Manchester. Therefore, their interests were linked.

The colliers from the Potteries, eight thousand strong, joined the Federation in January 1883. The North Staffordshire coalowners must have regarded this as a hostile act. In April 1883, they announced their decision to reduce wages by ten per cent. This was in a time when the prevailing national trend was for wages to rise. So the Staffordshire men thought they had a legitimate right to ask other colliers to help. But the M.N.U., with its no-strike policy, declined to assist, as did the various county organizations approached. Only Lancashire stood by them.

The Federation took its decision on 13 May, 1883. Members levied themselves ninepence a week initially and later sixpence a week to provide their Staffordshire brothers with ten shillings a week strike pay. Notwithstanding this aid, there were thousands of non-unionists in the Potteries. Unless they remained solidly on strike, the struggle would be lost. Yet, the Federation's limited resources made it impossible to provide for them as well.

The answer to the predicament came. The Federation issued a general appeal to all colliers in the North-West to support the Staffordshire men. A handbill, issued by Thomas Ashton on 12 June, 1883, explained why the union attached such great importance to the struggle:

> 'Shall we not buckle on our armour afresh and fight with great vigour? The battle at present is being fought in Staffordshire, but if we are beaten there, we shall have to either fight in Lancashire or surrender. The levy may be considered by some men very heavy, but the reduction will be threefold more. Who amongst our ranks would rather give the employers 2s. than the men 8d.? This and more you will have to do if the battle is lost. Let every union man pay his levy without grudging, and also get the non-union men to assist their comrades who are fighting their battle in the North Staffordshire district.'[7]

As a result of the appeal, the response was good. Collections were regularly made at Lancashire collieries. Non-unionists in Lancashire sent money to non-unionists in Staffordshire while the union members in Lancashire continued to support the union members in Staffordshire. This was a tremendous display of solidarity; according to Mr Page Arnot, unique in the annals of mining trade unionism.[8] Nevertheless, the Staffordshire miners lost. After a twenty-two week struggle, they were compelled to return to work, accepting a ten per cent cut.

The colliers' organization in the Potteries emerged from the contest severely debilitated. It remained in the Federation, growing weaker and weaker, finding it increasingly difficult to meet its obligations. In September 1886, the Federation decided to disaffiliate North Staffordshire because, over a prolonged period, it had failed to pay its levies.[9]

This collapse of a union, a union whose membership was higher and proportion of total labour force greater than in Lancashire and Cheshire, profoundly shocked the colliers of the North-West. It was an omen of what the future might hold for them—unless they could drastically alter the configuration of forces. The danger could only be averted through a search for agreement. This had to be conducted on two planes. First, there was the need for greater unity among the miners throughout the country. The creation of a unified organization, with a national strategy and tactics, would be a source of great strength. Second, there was the need to arrive at some *detente* with the employers, an agreement, however onerous, would be preferable to suffering the same fate as the North Staffordshire men.

The quest for agreement among the various colliers' unions was a slow, disheartening business. In spite of this, Lancashire saw it as vitally necessary. The Federation, therefore, took a leading part in convening conferences, endeavouring to foster a co-operative attitude between leaders and men in the various coalfields. As early as October 1882, Lancashire convened a national conference. The demand for a wage increase was synchronized in all the coalfields that were not bound by a sliding scale agreement. Each county union undertook to help any of the others that encountered opposition. Trade was at that time improving; the industry's earnings rising. Consequently, coalowners were prepared to concede a ten per cent increase and so no difficulties arose. Thomas Ashton believed that the attitude of the trade unions was shortsighted at that time. Instead of using the goodwill engendered by the wage increase to place their ad hoc national arrangements on a more enduring basis, each county organization simply slipped back into its own parochial ways.

Seven months later, the Staffordshire struggle showed how foolish this was. Thomas Ashton had predicted that the defeat in the Potteries would be a prelude to cuts in Lancashire. He was right. But the Federation attempted to continue pushing up wages. A ten per cent claim was submitted in the autumn of 1883. Yet, despite improvements in trade, the men received a blunt refusal. The Lancashire leadership then tried to repeat the tactic which had been successful the previous year, calling a special national conference at Sheffield on 3 December, 1883, to co-ordinate wages policy in all the coalfields. But no agreement could be reached. So a second conference was held at Manchester on 27 December, 1883. This revealed that more and more county organizations were having misgivings about making a concerted attack at that time. 'The delegates were of the opinion to enter a strike with forces at their command would mean failure,' wrote Ashton.[10] Nevertheless, those who favoured a unified policy still persisted, hoping to change their colleagues' opinions. Another national conference was held—at Birmingham on 24 January, 1884—making three in eight weeks. But this even more emphatically rejected the idea of united action: 'We cannot agree to take immediate general action on the wages question'; each county union was left to formulate its own strategy.

This decision left Lancashire colliers vulnerable. They had applied for an increase and been refused. They had appealed to other colliers for unity and been rejected. Their organizational weakness gave the coalowners an opportunity to seize the initiative and counter-attack. Through the spring and early summer of 1884, individual masters endeavoured to impose cuts. Disputes took place at Bickerstaffe Coal Co., Skelmersdale, Blackrod, Standish and Ashton-under-Lyne. To support those on strike, the Federation was compelled to levy its members still at work. The financial strain this imposed on the men, coupled with the success of most of the masters' attempts to inflict reduction, led coalowners throughout South-West Lancashire to think the time was propitious to introduce a general, all-round cut.

This was not in keeping with national trends. When the majority of Lancashire miners received notification of an impending reduction —27,710 out of the 27,952 represented at the Wigan pit-set conference of 12 July, 1884—they agreed to make a further attempt to enlist support from elsewhere. They wanted a national coal strike, a complete shutdown, to bring Lancashire coalowners to heel.[11] But the national conference declined to take such a portentous step. Instead, it offered to provide assistance for a struggle to be waged against one or two large firms. Lancashire again met to consider this proposal. It decided to recommend a fight to be conducted against the coalowners of Park Lane and Pemberton. The men there, however, refused to do battle and consequently the employers' terms were grudgingly accepted throughout the county.

The reason why the Park Lane and Pemberton miners refused to strike appears to be because they thought the probable outcome would worsen their position, making it even less attractive than if they accepted outright capitulation. Should they strike, the struggle was almost certain to be protracted. Consumers would be able to buy coal from elsewhere. The employers would be able to import blacklegs from Cornwall or Staffordshire, as had happened in the past. Also, as the months passed by and the struggle dragged on, it was likely to inflict lasting injury on the miners in those areas affected. Consumers might become accustomed to buying their coal from other places—indeed, enter into long-term contracts—so that, when the men ultimately returned to work, they should discover trade slack and part-time working prevailing. This was what had occurred in Lancashire as a whole after the 1881 dispute: the men returned to find only half-time working in most collieries; much of the Lancashire market had been seized by outsiders. Naturally, the Park Lane and Pemberton men did not want to suffer in isolation what had been the painful experience of the whole Lancashire colliers three years previously.

The lessons of 1881 had not been forgotten. *The Lancashire Miners' Federation Journal* elaborated on them when it was discussing the probable consequences of a third of British miners stopping work while the other two-thirds, either organized or with sliding scale agreements, continued to labour:

> 'If one hundred and fifty thousand miners cease work, what will it avail if another three hundred thousand who remain at work commence working half as much more time than they would if the other hundred and fifty thousand were at work? Lancashire has not yet forgotten the beginning of 1881 when it was computed that at least forty thousand miners in Lancashire were out on strike, and everything in the miners' favour for gaining the day, and when it was thought the coal owners were just on the point of yielding, although one of the greatest snow storms on record was raging at the time, thousands of tons of coal came rolling into Lancashire every day from Yorkshire, Derbyshire, Nottinghamshire, North Wales and other places, and not only did these outside districts supply the Lancashire markets during the strike, but they made long contracts in some cases with customers close to the pits from where they had formerly been supplied ... When the strike was over, certain collieries in Lancashire and Cheshire could not work for more than half time, although in the very depth of winter.'[12]

With such an experience as this fresh in their memory, it can easily be understood why small groups of colliers, like those of Park Lane and Pemberton, were unwilling to take unilateral action. Similarly, it is clear that Lancashire had strong reasons for favouring the formation of a national organization. A Federation conference in May, 1885, reiterated the view often given by Lancashire union spokesmen: 'Nothing but a general national movement can stop the continual reduction in miners' wages that are taking place.'

By mid-1885, the British coal industry as a whole was experiencing a recession. Widespread wage cuts made miners disgruntled and demoralized with their own organizations' failure to halt the downward plunge of wages. So when Lancashire took the lead in convening a national conference, hoping to lay the basis for united resistance, they met with little response. Thomas Ashton reported, 'The attendance was so poor, only delegates from six districts put in an appearance, that the president refused to open the conference.'[13] (It may well be that all six districts were from Lancashire.)

The obstacle preventing progress towards greater unity was that negotiations, in their entirety, were conducted at a local level. If national negotiations could be established, then miners might widen their mental horizons and the prospects improve for forming a national union. Clearly seeing this, Thomas Ashton made repeated approaches to the Mining Association, the body that represented the coalowners of Great Britain. He suggested it might be mutually advantageous to discuss pay and conditions on a national basis: 'Good might have resulted from such a meeting, and it might have brought closer identity between capital and labour.'[14] But each overture met with rebuff. The Mining Association refused even to hold discussion. Its secretary, Maskell Pease said, 'the wages question is outside the scope and province' of the employers' organization.

With all negotiations taking place locally, it was extremely difficult —almost impossible—to get the separate county unions to agree to a common policy. In some coalfields, such as Durham, Northumberland and South Wales, the miners had agreed to a sliding scale, which related wages to the price of coal and precluded strike action. So a sizeable segment of miners were, as a consequence, debarred from participating in joint activity over wages. Elsewhere, in other coalfields, the impediments to the development of a unified policy were considerable. Colliers tended to be preoccupied with relations at local level, the arena in which wage settlements were reached. What happened there appeared more pressing, urgent and vital than any national conference. Besides being more remote than the face-to-face confrontations they had with their local coalowners, national conferences frequently seemed to indulge in meaningless chatter, adopting resolutions that were mere words because there were no means to achieve their implementation. In these circumstances, it is hardly surprising that when the Lancashire and Cheshire Federation made one of its many appeals for greater unity, it should receive the typical reply: miners 'should devote their whole strength to building up a strong local organisation in each district first before attempting to establish a national organisation'.[15]

The failure of its efforts to create a national union left the Federation in an exposed position. Weak and virtually defenceless, the Federation realized the need to pursue its second tactic—an attempt to reach an agreement with Lancashire and Cheshire coalowners. The union leadership had never regarded these as mutually exclusive options—either a strong national organization or agreements with local coalowners—but had pursued both courses at the same time. However, once it was obvious that no national organization would arise in the immediate future and there would be no united effort to arrest the catastrophic fall in wages, then it gave added urgency to arrive at some sort of workable arrangement with the masters.

To the union leaders, the Federation appeared to be melting away beneath its feet. Membership dropped from 16,526 in July 1884, to 11,500 in September 1884, and then dropping to 7,968 by May 1886. The financial position became equally precarious. The statement of accounts for 1886 showed the Federation's income for the first six months was only £110, 10s. 2½d., while its expenditure was £159 9s. 1d.; for the second six months income was £108 11s. 11½d. and expenditure £101 16s. 10d.[16] The Federation was barely paying its way. Any large call on resources might easily result in complete collapse.

Daily the Federation displayed its weakness. An attempt to resist wage cuts at Skelmersdale resulted in mass sackings, the dismissal of five hundred men. At Carr's colliery, Prescot, a strike for the same reason led to defeat, with the resumption of work on condition that the checkweighman, John Fegan, be not re-engaged. He searched for work in St Helens and for miles around. No manager was prepared to engage him, and so he decided to emigrate to Australia. But neither Fegan nor the Federation could pay the fare—collieries had to have a special collection for him.

The union leadership realized that it must lessen the number of stoppages if the financial position were to be improved and its strength increased. For that reason, the introduction of a sliding scale seemed appealing. In 1880, Alexander MacDonald said he 'was not long engaged in these arbitrations when he saw that they were costly and cumbersome, and tended to continue rather than ally irritation'. The introduction of a sliding scale was a 'nobler and a better way' to resolve differences.[17]

The coalowners were strong advocates of such a scheme. They

recalled that the union had agreed to a sliding scale in January 1881; only its unpopularity with the men had prevented it working. As the bitter battle of 1881 receded in the memory, coalowners hoped a new attitude would arise among the miners. So they repeatedly called for these new arrangements. Ultimately, on 9 June, 1886, a joint meeting was held. For the employers, Hewlett outlined the advantages of having a sliding scale. The coalowners, he said, were anxious to avoid the unpleasantness of giving notice of wage cuts. Under a sliding scale, wage adjustments would occur automatically. Thus, disputes would be avoided, continuous output maintained for longer periods and trade would be kept in the county.

Although not mentioned by Hewlett, the employers had another reason for wanting a sliding scale—the benefits arising from lower costs would all accrue to them. This may, initially, appear to be a doubtful contention: coal mining, being an extractive industry, is subject to an inherent tendency to diminishing returns. As the coal near the surface was exhausted and it became necessary to dig deeper and deeper, then costs would tend to rise. However, this overlooks the countervailing effect of technical improvements. As Professor A. J. Taylor says:

> 'In coal mining, the role of the innovator is fundamental. Whereas in manufacturing industry invention has increased productivity and reduced costs, in coal mining it has been and is the indispensable condition of the industry's continued existence. Without the steam-engine, first for purposes of mine-drainage and then for the raising of coal from increasing depths, coal mining could hardly have survived into the nineteenth century: without improved system of ventilation and underground haulage its development after 1850 would inevitably have been severely limited. But, although the basic spur of technological action may have been the elemental one of survival, it is clear that it also carried implications for the trend of costs and of labour productivity. The upward or downward movement of productivity in coal mining may be regarded basically as the resultant of the opposing forces of the natural circumstances on the industry's existence on the one hand and the ingenuity of the mining engineer on the other: in Jevon's words, it is "a fair fight with difficulties".'[18]

Professor Taylor contends that, during the second half of the nineteenth century, there was a significant tendency for the productivity of labour to increase. This was principally due to improved conveyancing, which lessened the number of men employed on haulage.[19] He says,

'Until the 1880s productivity would appear to have taken an upward course; thereafter it fell steadily downwards.'[20] An article in the journal *Nineteenth Century* (December 1884) corroborates Professor Taylor's statement with the following figures:

Labour force	*Tonnage*	
514,933	163,737,327	1883
538,829	140,713,832	1874

In other words, in nine years the labour force fell by 23,896 while coal production rose 23,023,495 tons. Production per man was 261 tons in 1874; it rose to 334 tons by 1880.[21]

It was precisely because the employers would derive benefits from lower costs that the Webbs were opposed in principle to the sliding scale: 'There seems no valid reason why the wage-earner should voluntarily put himself in a position in which every improvement in productivity methods ... should automatically cause a shrinking of his wages.'[22] The Webbs went on to suggest that such arrangements made the workman 'a humble partner ... a partner without in particular, any voice in that adjustment of production to the intensity of demand upon which the selling price of his product, and therefore his livelihood necessarily depended'.[23]

In South Wales colliers had worked under a sliding scale since the debacle of 1875. Whenever a dispute arose over an agreement, a union official would be compelled to remonstrate with his men and demand its implementation. Consequently, trade union leaders, as Jevons points out, 'became very largely a servant of two masters'.[24] The equivocal role of the miners' organization and its leadership had a disastrous effect on the membership. 'Under these conditions effective trade unionism in South Wales disappeared', Page Arnot says.[25]

The sad experiences in South Wales did not influence the Lancashire men. In October 1884, the Federation affirmed its willingness to accept a sliding scale if the coalowners restored the last wage cut. The masters refused. But this did not end the matter. A special conference on 25 May, 1886, reaffirmed its preparedness to accept a sliding scale. It suggested that a variation of two and a half per cent in wages for a 3d. variation in the selling price. Negotiations were opened. Then doubts began to proliferate. The Hindley and St Helens colliers were afraid that to agree to a sliding scale without provision for a minimum beneath which wages would not be permitted to fall would be, in the prevailing state of trade, a reckless undertaking. Other miners objected to the owners' proposal for the base period, the time from which all calculations would be made. The masters, naturally, wanted the period that was most advantageous to themselves when wages reached their lowest point relative to coal prices. They refused to accept the Federation's suggestion of alternative dates. They also would not countenance the comparison of Lancashire wages with those in other coalfields before determining the base wage.

From the statistics contained in Table 14, it may be wrongly assumed that the Lancashire and Cheshire miners were comparatively affluent. But these figures conceal some important points. Much of the work in the North-West was done by daywagemen, colliers less well paid than those on piece-work. Also, Lancashire colliers were expected to pay from their own pockets for various extraneous expenses—ripping, timbering, etc.—which in other areas would be paid for by the employers. Therefore, the figures in Table 14 are not comparable. When allowance is made for this fact, then a different picture emerges. In his book *Wages in the Coal Industry*, J. W. F. Rowe concludes that the Lancashire collier's net pay was less than that of colliers elsewhere.[26]

The union became increasingly uneasy about the negotiations. It feared that, even if the sliding scale did determine gross wages, it would not prevent owners from nibbling away, reducing the actual take-home pay by making the miner responsible for more of the expenses. Its anxiety was not allayed when the employers refused to agree to satisfactory arbitration procedures for deciding this kind of contentious point.

The negotiations reached a dead-end almost by accident. In June 1887, another ten per cent wage cut was imposed. But Wigan Coal & Iron Co. decided to reduce wages from between seven and a half and twenty per cent at its Prospect, Taylor and John pits. Regarding the size of these reductions as unreasonable, the Federation agreed to support the men's withdrawal of labour. The Company responded by simply transferring work to its other collieries, places unaffected by the dispute. By October 1887, the Federation was counselling the men to submit, all further resistance being considered futile. Sam Woods wrote to A. Hewlett, the Company's manager, asking to open discussion for a return to work. But Hewlett refused to meet Woods even after the men had agreed to accept the management's terms *in toto*. Hewlett stated that the three pits would remain closed: the miners had to be taught a lesson.

The clear, unashamed hostility Hewlett had shown during this dispute was of the utmost significance. For he was the employers' main spokesman in the sliding scale negotiations, a fact that soured the entire discussions. Perhaps, from the miners' viewpoint, the final indignity came when a pit-set conference approached employers for discussions over deductions. During that period, according to Thomas Ashton, 'the system of deductions, especially where the miners have had no voice in the arrangements, were no less than barefaced robberies'.[27] Feelings were running high among the miners on this issue. Union records contain many references to 'the wholesale confiscation of the men's tubs'.[28] For instance, a hundred men who worked in the Orrell five-foot mine had three thousand boxes confiscated in two years.[29] Yet, the employers just brushed off the pit-set conference's suggestion, refusing to discuss the matter of deductions with the men.

Such intransigence from the coalowners, particularly the large coalowners, ruined all chances of getting a sliding scale introduced. In any case, elsewhere, the tide was moving against this kind of agreement. Sliding scales agreements were ended in Cumberland in 1881, Cannock in 1883, Northumberland in 1886, Lanark and Durham in 1889.[30] Lancashire had attempted to join the party when it was almost over.

With the end of negotiations, the Federation's search for agreement appeared to have failed on both fronts—it had not achieved unity with fellow miners in other coalfields; it had not come to an understanding with the employers. The men suffered in consequence. In 1887-8, the trough of the depression, short-time working was common. Wages fell to between 10s. and 18s. a week. Aspinwall said there was greater hardship than he had ever known. He told employers, 'Poverty, gentlemen, is rampant in the families of your workmen.'[31]

Fortunately, by the autumn of 1888, the coal trade generally was experiencing some improvement. As frequently happens during the upswing of the business cycle, the frustration of workers, latent and unexpressed in the hardest times, exploded with a new-found militancy. Miners forgot their objections to united action, which had seemed so weighty a few months earlier. When Thomas Ashton dispatched letters calling a national conference, his circular was received with enthusiasm. The delegates who gathered at Birmingham on 11 September, 1888 had a different spirit, a more confident outlook, than those who met at the last national conference, held in January 1888, which broke up through disagreement. This time Thomas Ashton, appointed secretary by the conference, told delegates, 'We are just bordering on a great battle, which will require all the forces under our command to win. There must be no division in the camp, one united effort of Lancashire, Yorkshire, Derbyshire, Nottingham, Staffordshire and North Wales will obtain and secure an advance in wages.'[32]

Thomas Knowles opposed the Employers' Liability Act. He told Parliament it would give workers an incentive to injure themselves deliberately. In this cartoon, he is seen talking to Lord Lindsay, his fellow Tory M.P. for Wigan.

The Birmingham conference agreed with Ashton's prognosis. It resolved to make application for pay increases in every coalfield. Wherever employers resisted, the miners would concentrate their combined strength.

A recall conference on 29 October, 1888, heard that in most places the coalowners had acceded without a struggle; only in Yorkshire, where thirty thousand were on strike, had difficulties arisen. Delegates decided to levy members a shilling a week—two shillings if necessary—to crush this resistance. When Yorkshire coalowners realized how determined the miners were, they capitulated, thus providing a useful example of the power of united action for advocates of a national organization.

Success breeds success: on 16 February, 1889, Lancashire approached other districts with the aim of repeating the exercise. The national conference was again held in Birmingham (26-29 March, 1889), and it again resolved to call for a ten per cent rise. This time, however, it met with stiffer resistance. Also, some owners proposed compromises which would have weakened the general cohesion. Yorkshire masters, for example, wanted to grant five per cent and a further five per cent at some future date. Employers in the Forest of Dean and Somerset had similar ideas.

The danger of fragmentation, a growing diversity of miners' standards, would sow the seeds of discord. Only if miners put their alliance on a more enduring basis could they prevent the masters' divisive policies tearing them apart. The slogan 'Unity is strength' had been proved, at least in a negative way—that is, disunity is weakness—in the 1880s. Thomas Ashton drew the lessons from this period when arguing the case for a national amalgamation:

> 'They remembered that in 1882, after a successful agitation, the men obtained an advance in wages, but there was no general or national organisation to protect and keep that advance secure. The representatives after the conference went back to their districts and left each district to do the best for itself, with the result that Staffordshire was attacked in 1883, Lancashire in 1884 and Yorkshire and other counties in 1885, and the whole of the advance obtained in 1882 was all gone and for several years the colliery workmen were in the most deplorable conditions.'[33]

A repetition of these calamities could only be avoided if district unions could not, when decisions went against them, just opt out. Therefore, it was resolved to form a national organization—the Miners' Federation of Great Britain. A small committee drafted the rules on

11 October, 1889. They were considered at the Newport conference a month later. Finally, the new union was born at the Birmingham conference, 23-24 January, 1890.

From the formation of the M.F.G.B., no counties benefited more than Lancashire and Cheshire. It transformed the relationship of forces throughout the coalfields of the North-West. Not only did it give an excellent opportunity to advance wages rapidly, it also led Lancashire coalowners to treat the union with greater circumspection. The union, with its increased power, had become a formidable adversary. Consequently, the union found that in many situations where previously the coalowners had been prepared to risk a strike, they now preferred negotiations.[34]

By the beginning of the 1890s, the Lancashire and Cheshire Federation could reasonably claim to have partially attained some of its objectives. The national union had been created, put on a lasting basis, never to be smashed. At the same time, Lancashire coalowners were becoming, in general, more amenable to persuasion, more ready to substitute discussion for dispute.

Both these objectives had been achieved partly through the slow, relentless and often seemingly unproductive toil of the union and its officers. But there can be no doubt that as well favourable economic developments helped to provide the conditions in which the work of Ashton and others could reach fruition. In the search for agreement they may not have arrived at their ultimate goal—at least it was clear that they were going in the right direction.

Table 14

Miners' Wages, hours of work and Union Membership in 1886

	Hours	*Pay per day*	*Membership*
Scotland	11-12	2s.—3s. 6d.	25,500
Northumberland		5s. 2d.	13,000
Durham		4s. 4½d.—4s. 7½d.	30,000
Cleveland		4s.—4s. 2d.	
Derbyshire	8½		2,500
Lancashire	9½	4s. 9d.	7,000
S. Staffs.	8	3s. 9d.	
N. Staffs.	8	4s.	
Cheshire	8	3s. 9d.	
Cumberland	8	4s. 6d.	1,000
Nottingham	9	4s. 9d.—5s. 3d.	500
Aberdare	9	3s. 10d.	6,200
Monmouth	9	3s. 4d.—3s. 10d.	
N. Wales	54/week		9,000

Figures given to the Manchester conference, 28 January, 1886.

CHAPTER TWELVE

The End of the Old District Unions

The formation of the M.F.G.B. coincided with prosperity in the coal trade. The quantity of coal sold and the price at which it was sold rose rapidly. So did wages. By June 1890 they reached their highest-ever point, 40 per cent above the 1888 level.[1] The swift accretion of wealth made miners favourably disposed to the new enterprise. It gave the M.F.G.B. a beneficient blessing, helping its leaders to place the new-found unity on an enduring basis. Colliers tended to attribute their personal prosperity to their organization rather than to the natural workings of market forces. In fact, it was partly both. Wages would have tended to rise anyway. But possessing a strong centralized national leadership, powerful enough to wring concessions, colliers belonging to the M.F.G.B. could fully exploit their position. However, in those coalfields where men remained in independent county organizations, affiliated to the loose, federal M.N.U., wages limped behind prevailing market trends.

The two national bodies did not merely have different structures but also outlooks. In 1892 the M.F.G.B., anxious to run down the large amounts of coal accumulating at pit-heads, resolved to have a nationwide one-week stoppage. But the M.N.U., averse to all forms of industrial action, did not take part. So the M.F.G.B. secretary, Thomas Ashton, appealed to the miners outside its jurisdiction. He publicly asked them, 'Are the men in these districts going to work overtime whilst our men are playing?' For miners to continue piling up coal elsewhere while the M.F.G.B. were on strike was self-defeating a point that produced great annoyance.

Relations between the two national bodies grew acrimonious. Thomas Burt, under pressure from his members to explain why wages in other places were rising faster than in the North-East coalfields, argued that the newly-formed M.F.G.B. was not responsible. It was entirely due to the pressure of demand in different parts of the country, regional fluctuations in market forces. To which Thomas Ashton retorted, 'His views and ideas are too old and musty for the present time.'[2]

Soon after the M.F.G.B. had held its national week's stoppage, the

Durham miners, who had not participated, found themselves confronted with a wage cut. On their own, they were powerless to prevent it. For the M.F.G.B., Thomas Ashton could not resist the temptation to point out the lessons. He reminded the Durham men that of what he had said a few months before: it was a dangerous practice to leave large quantities of coal at the pit-top, an incitement to owners to reduce wages. Also, he claimed that the Durham miners enfeebled themselves by belonging to the M.N.U., which would never provide assistance, instead of the M.F.G.B.:

> 'The Federation policy may be called noisy and aggressive, the National Union watchful, firm and conciliatory; but the aggressive policy is what the men require. Of what use is the firm and watchful policy of the National Union to the poor, struggling men of Durham? The men are simply being starved into submission. Oh, for one month of the aggressive policy of the Federation! Had Durham been members of the Federation, a general strike would have been declared, and none would be watching, all would be fighting. There is too much watching and too little work.'[3]

The great test of the relative strength of the M.N.U. and the M.F.G.B. came during the 1893 lock-out. In the Federated districts, coalowners became irritated at the widening disparity between the wages of their miners and those who were members of the National Union. A dip in the level of industrial activity provided them with a suitable opportunity, they thought, for bringing the pay of their employees back into line. They issued an ultimatum to the M.F.G.B.: accept a 25 per cent drop or the collieries are closed.

Obviously, the coalowners did not anticipate the unity and tenacity of the miners, determined to resist any cut whatsoever. There was virtually a 100 per cent stoppage throughout the coalfields. Suddenly, coal supplies were shut off from all M.F.G.B. districts. Immediately prices started to rise, and continued to do so at an ever-increasing rate. Some owners capitulated, re-opening their pits to gain advantage from the temporarily buoyant market. This led others to follow suit. Until, after 16 weeks, those owners still struggling on saw its futility. The men had won.

The 1893 lock-out had involved three hundred thousand miners. Twenty-one million working days had been lost. According to Messrs Clegg, Fox and Thompson, it was the biggest trade dispute this country had so far experienced.[4]

The significance of the outcome is hard to exaggerate. Affluence faced poverty in the mining communities. Wage cuts had successfully been imposed in Northumberland and Durham, the men of South Wales were working at rates 12½ per cent below those of 1879. In contrast, the miners of the M.F.G.B. were still working at 40 per cent above the 1888 pay rates. Colliers in the Federated areas could see for themselves, in tangible monetary terms, the superiority of the new form of organization.

Also, a precedent had been set. For the first time in history, miners' wages had failed to fall with the fall in coal prices. Hitherto, unions that had opposed the downward trend of pay rates had been smashed. This time they emerged strengthened and triumphant.

This gave meaning to the concept of 'a living wage'. There should be, it was argued, a certain figure below which miners' incomes should not be allowed to fall. This should be determined, not by the vagaries of market prices, but by the amount required to live with a reasonable standard of food and comfort. Sam Woods explained the principle:

> 'Notwithstanding all the teachings of political economists, all the doctrines taught by way of supply and demand, we say there is a greater doctrine overriding all those, and that is the doctrine of humanity. We believe, that the working man is worthy of his hire, and hold at the present moment that wages are as low as they ever ought to be.'[5]

Thomas Ashton concurred. In retrospect, he wrote, 'The successful termination of the great strike of 1893 was the means of uplifting the mineworkers throughout Great Britain to a much higher standard of life, and permanently placing them in a much better and more respectable position in society.'[6]

All this had been accomplished, as Thomas Ashton never tired of saying, through the power of organization. When opening the Tyldesley Miners' Hall in 1894, he said the lock-out had been 'won principally through the grand discipline of the Miners' Federation of Great Britain'.[7]

Paradoxically, while Thomas Ashton could refer proudly to the prowess of the M.F.G.B., the same language would be inappropriate to describe the position in the North-West. Indeed, the strength of the national Federation only served to underline the weakness of the local Federation. This had been clear from the day the M.F.G.B. was formed. Much of the preliminary work had been done by Thomas Ashton. Yet, ironically, he could not persuade the Lancashire and

Cheshire Federation to send delegates to the inaugural conferences. Mindful that the Federation had not been represented at the 1889 T.U.C., Thomas Ashton dispatched a letter to all lodges, pleading with them to reconsider their decision:

> 'The various Mining Counties in Great Britain will be represented, except Lancashire. As the last Lancashire Federation Conference decided not to appoint any delegates, but to leave the matter with Districts, I trust that the large Districts will appoint Delegates, or the next Federation conference will rescind the resolution.'[8]

In the event, Lancashire was over, not under, represented. Each district union felt obliged to send its own delegates. Rightly regarding the M.F.G.B. as a potentially powerful body, district unions were too jealous of their own prestige to allow some other organization—the Lancashire Federation—to represent them. Hence, whereas small and united delegations came from other coalfields, Lancashire's delegation was usually distinguished by its size and internal squabbles. It consequently did not have a coherent, powerful voice in national decision-making.

This is directly attributable to that peculiar characteristic of miners' organizations in Lancashire—the district unions. As already stated in Chapter Three, these probably arose partly because of the lack of any accepted geographical focus. In most other coalfields, there was a definite town—Newcastle in Northumberland, for example—where county meetings were always held. But Lancashire had no such place. Meetings were usually held in Manchester, Bolton or Wigan, while the owners preferred Liverpool for their gatherings.

This absence of a clear-cut centre created conditions favourable to polycentrism, the development of power in the various localities. The Lancashire strike tactic helped to further this process. One area would stop working and be supported by all those remaining at work elsewhere as well as non-miners in the dispute-affected town. Help from other sections of the working class came because of Lancashire's variegated economy, where people earning their living in different ways lived in close proximity or were even members of the same family. Naturally, with such close ties, people automatically helped one another in times of difficulty.

The Lancashire strike tactic was a kind of guerilla-styled industrial warfare. Because of the localized character of each contest, groups of workers were, in an important sense, dependent on their own initiative. They took the decisions on matters of consequence, which largely determined the ultimate outcome. For this kind of struggle, a central leadership, giving orders, formulating policy, was an irrelevance.

Both the strength and weakness of district unions came from their intimate links with the social life of the localities: strength since their ties to the local community made them an integral part of its being; weakness since the community's corporate spirit involved proudly distinguishing itself from neighbouring communities. Union meetings, usually held on licensed premises, generally took place in a friendly, convivial atmosphere. In Leigh, for example, there were 19 separate lodges meeting in 19 different public houses.[8a] Colliers, free to belong to whichever lodge took their fancy, customarily joined the same one as their friends. While these arrangements enriched miners' social life, as well as having a beneficial effect on the brewery trade, their effect on trade unionism was catastrophic. With men from a particular colliery attending numerous meetings, when a problem arose at that mine it might be independently discussed at lodges scattered throughout the town. As a result, half a dozen different and conflicting decisions might be taken. A further disadvantage was, as Thomas Ashton explained, 'Under this system, it was almost impossible to find out whether certain men were members of the union or not.'[9]

Consequently, the district union form of organization helped to keep the percentage of men in the union low as a proportion of the total labour force. Moreover, it weakened the whole fabric of the Federation by engendering rivalry and discord.

This arose because each district union saw itself as a competitor with other district unions. It behaved like a typical small business. It aimed to attract custom from its competitors by either lower subscriptions or higher benefits. In a way similar to a capitalist entrepreneur, a district union would strive to extend its range of operations. This biggest district in the coalfield—Ashton and Haydock—appears, like Topsy, just to have grown and grown. For no apparent reason, it included Burnley in its area.

As their was no clear principle to determine to which district union a branch should belong, disputes inevitably occurred. They were particularly rife when a new colliery opened, an occasion on which many unions might vie for custom. This generated mutual recriminations that could spread to other contentious issues. A typical instance is contained in the Lancashire Miners' Federation minutes of 2 June, 1894:

'An objection was raised by Tyldesley, Leigh and Atherton against a new lodge joining the Federation through Platt Bridge District. The lodge was split from Leigh.

'Platt Bridge then raised an objection against a Lodge which had joined the Federation through Bamfurlong, which had been opened at a school-room, one portion of the members previously belonging to Platt Bridge. Bamfurlong then raised an objection against Ashton and Haydock taking their members.'

Some small district unions had no conception of what was required to run a large, efficient county Federation. The problem of developing a leadership, men with sufficient knowledge and skill to tackle the complexities that arose, was beyond their comprehension. Blackrod believed that all offices in the Lancashire and Cheshire Federation should be filled by drawing lots.[10] Standish believed that places on delegations should be taken in rotation.

Not surprisingly, the 1893 lock-out, which vindicated the organizational form of the M.F.G.B., also tested that of the Lancashire and Cheshire Federation and found it wanting. Since it was difficult to say who were members of the districts, it was difficult to decide who were entitled to strike benefit—or, for that matter, those at work who should be contributing to the levies. To compound the confusion in the middle of the lock-out, Ashton and Haydock seeing the muddle, proposed that the Lancashire Federation be reorganized.[11] Of course, it was impractical at that juncture. Nevertheless, it was extremely significant. The biggest district, with some of the most influential members in the Federation, realized that the chaos must cease.

As if to confirm this view, the Oldham district collapsed at the end of the lock-out.

The Ashton and Haydock proposal would have tended to strengthen larger districts at the expense of smaller ones. It would have produced a rationalization of a federal structure, not the creation of a centralized one. The smaller unions (those with less than 200 members) were to be disbanded; unions that refused would receive no further help from the Federation. The most important suggestion, that dues be paid at the pit-head and that miners at a colliery should all belong to the same lodge, would have lessened confusion considerably.

But after the lock-out most miners were in no mood to agree to such sweeping changes. It was not merely that they threatened the vested interests of smaller unions. Their introduction would necessarily have led to a first-rate quarrel. The Federation could ill-afford the luxury of the heightened internal dissension that it was certain to occasion: the external threats seemed so menacing. First, there was the Leigh Reform Association. This was a breakaway group, formed by colliers angry with their agent, R. Isherwood. He had simply endeavoured to create some semblance of order at Leigh. In a small way he had tried to do what Ashton and Haydock wanted throughout the North-West: for miners to belong to the lodge at the colliery where they worked.[12] He had, as well, wanted union meetings held in schoolrooms instead of public houses. If these proposals met with stiff resistance—mass resignations and the creation of a breakaway organization at Leigh—then it was surely reasonable to assume there would be a similar reaction, on a much larger scale, should they be applied in all Lancashire and Cheshire.

The second menace came from a second breakaway. The Conservative Labour Union was formed to protest against the candidatures of Aspinwall and Woods, the political involvement of the Federation with the Liberal Party. It constituted a serious danger because, unlike other coalfields, Lancashire had large numbers of miners with strong Conservative convictions. Men of the calibre of Thomas Oakes, the militant leader of the 1881 strike, belonged to 'the blue button union'. For a time, he persuaded his own district, Hindley, to withdraw from the Lancashire Federation. Elsewhere membership suffered from resignations made because of political disagreement.

Haunted by the spectre of irate colliers, some defending their rights to still have their traditional-style lodge meetings in public houses while others defended their Conservative beliefs, the Federation had second thoughts about reorganization. Toryism and beer is a powerful combination. No wonder union leaders drew back in fright!

But the problems remained. The Federation minutes contain cases such as that at Brynn colliery, Bamfurlong, where a strike occurred in January 1895. 'More union men from the Wigan district worked at this colliery than from the Bamfurlong district,' states the Federation report, 'and Wigan district had heard nothing of the dispute.'[13]

So, once the external threats had abated, the Ashton and Haydock district again raised the question of reorganization. The Federation agreed to set up a small sub-committee—Isherwood, Greenall and Aspinwall—to investigate the possibility and formulate proposals. They reported to the executive in April 1895, and it was agreed that the Federation structure be transformed, representation being on a branch rather than a district basis.[14] Even a full delegate conference accepted this commitment and passed the following resolution:

'That this conference is of the opinion that the best system to adopt to place this county on an equality of organisation with other mining counties in England is to centralise the management of the Federation, and that all the permanent officials or agents now paid by districts be paid and controlled by this Federation.'[15]

But, once this resolution had been passed, miners started having misgivings and second thoughts. Tyldesley, Leigh, Atherton and Hindley districts declared against it. Aspull, arguing for the *status quo*, said it thought they already possessed 'the best organisation that has ever been in operation'.[16] Ironically, Aspull had only joined the Federation, which it regarded as so magnificent, in 1889; a feud dating back to 1872 had kept it aloof from all county connections for 17 years![17]

Opposition to reorganization was deep-rooted. Union leaders tried to combat it by holding meetings throughout the mining districts. Two representatives of the Federation spoke at each one, explaining the advantages of the change. But, despite these efforts, the result of the ballot was, as Thomas Ashton said, 'not satisfactory. The question was left over for further consideration.'[18]

This setback could not be the final word. The pressure of events compelled them to return to the subject. This time it was agreed to forgo the luxury of a ballot. They merely drafted the new rules, submitted them to a special Federation conference, and gained its assent. The centralized structure came into operation on 9 January, 1897. As a sop to localist feelings, district unions were still permitted to continue to exist. But they had lost most of their real power. The Federation itself appointed miners' agents, delegates to national conferences, and had most of the wealth. It was also the supreme policy-making body.

A new ear had opened for trade unionism in the Lancashire coal industry.

In considering the years before centralization, Lancashire's organizational weaknesses must constantly be kept in mind. Localism, linked to law membership, created problems which, on that basis, could not be solved. Even in moments of elation, union leaders' thoughts were liable to be jogged back to these inadequacies. This happened in 1889. As we have seen, this began a period of rapidly rising wages. Lancashire owners granted these without much opposition. Negotiations remained cordial until Sam Woods raised the question of daywagemen's wages. For the employers, Hewlett gave a dusty answer: 'Oh, the daywagemen are our men; the colliers are yours.'[19] Consequently, increases conceded to colliers were either given in part to daywagemen or not at all.[20]

Customarily, colliers regarded daywagemen as inferior beings. 'You are only a pit topman' was a typical saying, with the implication that status rose the nearer one went to the coal-face.[21] Since daywagemen were held in low esteem, little effort was made to recruit them into the union. But this had to be altered if, as Sam Woods said, the union were 'to attack the fiend of disorganisation'. The Federation dispatched a questionnaire to discover how many daywagemen there were in various localities and how many belonged to the union. Then the Federation held a special recruiting campaign. But it does not seem to have been entirely successful. Large pools of unorganized labour remained.

In an endeavour to obviate the pernicious effect of operating an organization in an unorganized context, the union attempted to involve non-unionists in its activities. For instance, when the first miners' gala was held at Southport in 1890, a cordial invitation was extended to union and non-union miners alike. In a spectacular demonstration of solidarity a procession, estimated by *Labour Tribune* at between sixty and seventy thousand, marched along the beach and held a meeting near the pier.[22]

Likewise pit-set conferences were a device for stimulating non-union men to act in an organized manner. It gave greater cohesion to miners as a group. But this stratagem also had disadvantages. There was no way of ensuring that a pit-set conference's decisions were implemented. Little, if any, protection could be given to men who tried to carry them out and thus incurred the wrath of their employers. Also, pit-set conferences, where the poorly paid and most oppressed were more likely to be represented than at the usual union conferences, were liable at times to be more volatile, making militant decisions without having the financial resources to give effect to them. Since it would be the union's money that would be expended in any struggle, the Federation made a rule that a pit-set conference never had the authority to call a strike. Notwithstanding this, Ashton and Haydock believed that the practice of holding pit-set conferences should be discontinued as they tended to behave in irresponsible ways.[23] Others did not agree. They continued to be held until they became superfluous—that is to say, until the membership expanded to such an extent that it was almost the same as that represented by a pit-set conference.

Towards that end, the Federation arranged recruiting campaigns in specific places. One at a time, the districts would be granted permission 'to force the whole of the men in their district into the union'. What this involved was a concentrated attempt to persuade miners in a locality to join. If this failed in some instances, then the union would 'use force'—that is, tell their members not to work with non-unionists and to ostrasize them socially. Since this could lead to conflict, it was important that the majority of union members in the county remain at work, ready to give financial help. In effect, it was a variant of the Lancashire strike tactic.

By this method, the union was fairly successful in improving membership figures. The problem was that it failed to be efficacious in the really black spots, those where the union was virually non-existent. Big firms could respond to a piece-meal attempt, one confined to a single district, by sacking all known union members, closing the colliery if necessary, and transferring production to another part of their industrial empire.

Andrew Knowles and Son behaved in this way. As a result, it defeated attempts to unionize its labour force. Thomas Ashton explained, 'After the great lock out of all the men employed at A. Knowles and Son Ltd, in the year 1866, there had been no union in connection with the men employed at Pendlebury for over 20 years.'[24] In 1891, however, interest was rekindled when the Company imposed new working arrangements which were tantamount to a wage cut. It decided that colliers would have to do more packing. Instead of having daywagemen packing five yards in ten yards, the length was to be extended to 16 yards, with colliers packing the extra six on top of their customary five. One hundred and seventy-five men at the Wheatsheaf colliery, Pendlebury, stopped work and contacted the union.

The union strove to reach an understanding with Andrew Knowles and Son. Sam Woods wrote to Bramall and Knowles, the two managers, asking for a meeting, but received no reply. Meanwhile, the Wheatsheaf miners received, as well as normal strike pay, an extra 4s. 6d. from the Federation as an encouragement. At the same time, the union attempted to establish contact with the 3,000 other men employed by Andrew Knowles and Son. Failing a settlement, it hoped to have a general strike throughout the Company's collieries. But only half the men bothered to vote in a ballot—1,583 for, to 219 against—to decide whether to come out on strike. In view of this disappointing result, it was thought advisable to make more efforts to reach agreement with the firm.

Two and a half months after Woods' unanswered letter, Thomas Ashton wrote to Henry Bramall:

'I trust you will at this time prevail upon Mr. Knowles to break through the old conservative tenets, which belong to the last generation, and let the delegation from the Federation come and meet you, and in a friendly spirit settle the matter.'[25]

Eventually, Bramall and Knowles relented. Meetings were held to discuss what proportion of the packing should be done by colliers and how much should be left for the daywagemen. After much discussion, it was decided to return to the old 50-50 basis. But once agreement had been reached between the Company and the union, it was put to the men who rejected it. So Thomas Ashton had to write to Bramall again. This time he refused to meet union officials unless they had powers to reach a settlement. The Federation now stepped in, using its influence on the Wheatsheaf men, getting them to accept a 50-50 agreement.

This ended the strike. The union could claim a partial victory. Andrew Knowles and Son had at last been compelled to recognize the union's existence and negotiate with it. But recruitment among the Company's colliers was less than anticipated, primarily because the men there believed the ultimate outcome did not compensate them for the months of hardship.

These men exemplified a disquieting tendency the Federation came across elsewhere: when non-union men, who had been oppressed for a prolonged period, start indulging in trade union activities, they have an inclination to fly to the other extreme—militancy—and are not prepared to accept what union leaders regard as reasonable compromises. Moreover, once their tempers had cooled, they were liable to relapse into their previous apathetic state. This seems to be what happened at Andrew Knowles and Son.

Fourteen months later, W. T. Postlethwaite, a socialist schoolteacher, described by the *Workmen's Times* as 'one of the ablest speakers in the North of England', visited Pendlebury and tried to get the miners to join the Independent Labour Union. He described the effects of the butty system, the Midlands system of sub-contract labour, which Andrew Knowles and Son had introduced into Lancashire:

'Most of the mines seem to be held even now by the favourites, or "charter masters", whose wages will sometimes reach as high as £3 or £4 a week, while thousands of others, who do the real hard work, cannot get enough to keep them in the bare necessities of life.'[26]

In December 1894, Thomas Ashton approached Andrew Knowles and Co., asking it to abolish the butty system. In reply, Bramall said he had no intention of interfering with freely made arrangements between man and man. This answer dissatisfied the Federation. It condemned the butty system and resolved, if feelings were sufficiently strong among the men at Pendleton, to sanction a withdrawal of labour. But only 159 of the 400 working at the colliery voted for strike action, so the project was dropped. The butty system continued until the firm ultimately decided it was not the most efficient method.

Probably a contributory cause of the union's relative powerlessness when faced with large employers arose from the fact that the large coal companies frequently diversified their economic activities, thus repeating the pattern of diversity that characterized Lancashire industry as a whole. A balance sheet of the Executors of the Duke of Bridgewater's estates contains the following figures for profits:[27]

£73,252	cannel
£3,157	carrying
£9,029	Ellesmere Port
£18,365	Collieries
£19,237	estate
£123,040	

The Executors' diversity of income acted as a kind of insurance. An industrial dispute in one branch of its activities need not unduly affect the performance of the rest. Consequently, the Company was well-placed to withstand a prolonged onslaught from the miners' union.

The Lancashire Coalowners' Association, a body dominated by the large owners, contemplated industrial strife with equanimity. In its evidence to the 1892 Royal Commission on Labour, it admitted there were 'constantly little disputes going on'. But it did not seem particularly worried. Its spokeman rejected the idea that there should be a joint committee or conciliation board.[28]

The Federation, paradoxically, was plagued with the very small as well as the very big coalowners. The former, men with little capital, employing a handful of miners, were usually marginal enterprises, where the owners had to watch every penny. Frequently, the boss was a self-made man, proud of his precious mine, and intensely hostile to all trade unions. Running his concern on paternalistic lines, he thought the method that the workmen should adopt to increase their incomes was hard work, like he had done. The journal *Justice* interviewed just such an owner—H. W. Hobart, who owned the Facit colliery, previously belonging to John Lord and Son, of Oldham. He had started down the mine at the age of seven and still worked as a collier. When the interviewer asked, 'How did you become a coal-owner?' He replied, 'I were careful and saved t'brass and bought it.'[29]

Many small pits were in North-East Lancashire. The geology of the district made it highly suitable for the sinking of mines with shallow shafts or running a drift mine into the slopes of the Pennines.[30] Labour conditions were also propitious. Cheap, hard-working, non-unionized workers gravitated into the mining areas around Burnley, providing a plentiful supply of men and thus giving owners every opportunity to weed out 'troublemakers'.

Giving resilience to the small owners' opposition was a few large owners, equally vehement about keeping unions out. The result of this united opposition was that, ever since the demise of the Amalgamated Association in 1875, colliers in North-East Lancashire had remained unorganized.

A sign that times were changing came in 1889. One thousand one hundred men in the Burnley area joined the union. Initially, the masters replied with sackings—107 were dismissed.[31] But this does not appear to have halted recruitment. Union leaders hoped to win the confidence of the masters, arriving at the type of working arrangements that existed in many other parts of the county. As a proof of their goodwill, they agreed to leave one of the most contentious issues—the installation of weighing machines on the pit banks—in abeyance for two years, until July 1891.[32] But this appeasement proved futile. At the end of the period, the owners still remained as adamant as before. Moreover, the temporary recession in the coal trade made them anxious to exterminate all union influence.

The first dispute arose during the one-week nationwide stoppage, organized by the M.F.G.B. in 1892, to reduce coal stocks. In the Rossendale Valley only three of the eight pits heeded the call.[33] The owners maintained full production in all of them, employing daywagemen in place of the strikers. At the end of the week the principal owners, George Hargreaves and Co., refused to re-employ any union men, except for those with large families. Spontaneously, the pent-up ire of the colliers expressed itself. A strike wave spread throughout most parts of North-East Lancashire. Thomas Ashton reported, 'The men on strike refused to accept the advice of the leaders.'[34]

Disregarding the other owners, Ashton and Haydock, the district union responsible, decided to concentrate its efforts on George Hargreaves and Co. This was a good tactical move. A display of militancy allowed them to re-gain control of their rank and file. Also, it seemed likely that, if George Hargreaves and Co. were defeated, it would have a tremendous psychological effect on the other owners. So Ashton and Haydock decided that, unless the 75 men got their jobs back, they would call out all four hundred men who worked for the firm.

In June 1892, Sam Woods reported to the Federation that they intended to mount an all-out attack. The news received a divided reception. Many Federation officials expressed doubts about the wisdom of the move, believing arbitration was the wiser course. Nevertheless, the Federation eventually left the decision in Ashton and Haydock's hands, and all the men were brought out on strike. The men received an extra five shillings strike pay as an encouragement. Other districts were urged to give financial support, too: 'The men's organisation is in danger, and the fight is for liberty.'

To encourage strikebreakers, the employers increased their wages and gave an undertaking to continue with their services once the dispute was over. A wallposter read:

'ACCRINGTON COLLIERIES

To Miners employed at those collieries

In consequence of the annoyance and trouble which you are subject to by picketing and following now going on, we have determined to give a further 10 per cent advance, and we wish you to be completely and fully assured that no man now working for us will ever be asked to make room for any of the old men, no matter how long the strike may last.

Geo. Hargreaves and Co.

Accrington, 23 August, 1892.'

These inducements, eventually raised to 30 per cent, attracted labour from less well-paid workers in the area as well as from Cornwall and Scotland.[35] A Federation conference on 24 September, 1892, when the dispute was four-months-old, heard that the collieries were rapidly being filled with blacklegs. Attempts by Thomas Ashton to open negotiations were brushed aside. The employers answered that the collieries now had almost their full complement. They would be prepared to accept only 40 men back. Federation officials—Woods, Greenall and Glover—recommended accepting defeat and letting the strikers ballot for the 40 places. But one delegate replied, 'They came out in a body and they ought to go back in the same.'[36] Dewhurst, of Baxenden, pointed out the inconsistency of the union position: it had been eager to get them to strike; now it was eager to stop them.

Even so, the miners were compelled to recognize reality. By February 1893—that is, after eight months' dispute—the Federation was informed all the jobs were taken. With no possibility of gaining employment at Hargreaves again, the strikers were offered two weeks' benefit and advised to look elsewhere for work. Ashton and Haydock district protested at this decision. It was finally resolved to pay them until they found other employment. Many of them were still out of work in June 1894.[37]

Another twist to the tangled web of industrial relations at George Hargreaves and Co., came when Ashton and Haydock, accepting the position as it was—namely, that the Company had a fresh labour force—took the logical next step and wanted to persuade the miners at present working there to join the union. At first the Federation did not agree—indeed, Ashton and Haydock's motion did not get a seconder—but, at the following meeting, when couched in more general terms, it gained permission 'to force the whole of the men in their district into the Union'.[38]

This led to an ironical situation. Regarding the strike as over, George Hargreaves and Co. began taking off the 30 per cent bonus in 5 per cent stages. So their miners turned to the union and asked if it would support a strike! But the Federation reminded them that they had been blacklegs responsible for the failure of the previous strike, and refused to give support.[39]

Not surprisingly, the Federation regarded North-East Lancashire as a problem area where, unless extreme care and caution were exercised, it could easily bankrupt itself in futile disputes. Whenever possible, the Federation placed its influence on the side of moderation. Even this was fraught with difficulty. Both masters and men in collieries around Burnley always appeared determined to do battle. There seemed to be very few moderates around.

The whole area seemed to be infested with a disputes virus. At Cliviger a strike broke out over the size of pillars and the management's failure to introduce weighing machines. The following month the manager at Hapton Valley dismissed a committee man. The local lodge retaliated by getting the engineman not to wind non-union men. As a result, he was dismissed. Reporting the incident, John Leeming, of Burnley, asked the Federation to call a pit-set strike at Hapton Valley, but it declined to do so.

Each time the Federation failed to fight, it strengthened the hand of the militant wing. In North-East Lancashire, Marxist ideas had gained widespread credence among colliers, as will be shown in Chapter Thirteen. The main spokesman for the militants was John Sparling, a member of the Social Democratic Federation on Burnley Borough Council. In 1895, the militants accused Thomas Ashton of siding with the employers over a wage claim at Great Harewood—a charge he stoutly denied—and then unsuccessfully ran Sparling against Ashton for the secretaryship of the Lancashire Federation.[40] But the fact that Sparling had been nominated by the biggest district union—Ashton and Haydock—was symptomatic of the extensive criticism of the moderate leaders.

The essence of the miners' predicament in Lancashire was that they were trapped in a vicious circle. Their fragmentation and localism produced a weakness of organization; the weakness of organization made it impossible to deal adequately with problems, such as the daywagemen, the big coalowners, and the North-East part of the county; and the failure to deal effectively with these problems led to more disenchantment and a further weakening of the organization.

The 1893 lock-out vividly illustrated this point. At that time, Lancashire and Cheshire was the second largest coal district in Britain. It employed 84,016 men and produced 23,022,944 tons of coal a year. Yet, Federation had £214 in the bank. Without outside help—in the last five weeks of the lock-out the M.F.G.B. alone sent £26,322—it would have been impossible to sustain the strikers.

The need for organizational changes in Lancashire was given added urgency by the changes made by the masters. They formed a Coalowners' Association Defence Fund. It had played a vital role at George Hargreaves and Co. For the union, the danger was that, backed by this Association, the coalowners would systematically wipe out the union, colliery by colliery. Only similar—or, better still, superior—organization on the miners' side could avert this calamity.

Table 15 *Production costs of the Arley Mine*

	Half Year ending 30 June, 1880	Half Year ending 31 Dec., 1881	Half Year ending 31 Dec., 1889
average selling price per ton	4s. 11·3d.	5s. 6·2d.	6s. 5·1d.
underground wages per ton	2s. 6·8d.	2s. 7·7d. }	3s. 5·8d.
surface wages per ton	4·4d.	4·4d. }	
royalty	6·5d.	6·4d.	6·5d.
other payment—materials, salaries, etc.	2s. 6·4d.	1s. 10·7d.	1s. 8·9d.
Total cost of production	5s. 6·6d.	5s. 5·2d.	5s. 9·2d.
etc.	2s. 6·4d.	5s. 5·2d.	5s. 9·2d.

Production costs of Yard Mine

	Half Year ending 30 June, 1880	Half Year ending 31 Dec., 1881	Half Year ending 31 Dec., 1889
average selling price per ton	4s. 11·8d.	5s. 1·4d.	6s. 7·5d.
underground wages per ton	2s. 9d.	2s. 9·1d. }	3s. 8·8d.
surface wages per ton	4·4d.	4·4d. }	
royalty per ton	6·2d.	6·0d.	6·3d.
other payments—materials, etc.	1s. 11d.	1s. 9·5d.	1s. 6·6d.
Cost of production	5s. 6·6d.	5s. 3·2d.	5s. 9·7d.

Source: Alfred Hewlett, of Wigan Coal and Iron Company, gave the statistics in his testimony to the R.C. on Mining Royalties.

Table 16 *Lancashire Miners' Wages, 1886*

	Weekly Wage	Days per Week	Wages per Shift
Hewers:			
Piece	25s. 0d.	5	5s. 0d.
Day	n.a.	5	n.a.
Putters	21s. 8d.	5·1	4s. 3d.
Stonemen	25s. 9d.	5·1	5s. 1d.
Firemen	28s. 7d.	6	4s. 9d.
Others	21s. 5d.	6	3s. 7d.

Source: 1886 Wage Census.

Daily Wages in West Lancashire, 1891

Colliers	7s.	0d.
Drawers	5s.	6d.
Datallers (skilled)	5s.	2d.
(unskilled)	4s.	0d.
Firemen (weekly	33s.-36s.	

Source: Fourth Report on Trade Unions, 1892.

CHAPTER THIRTEEN

The Struggle for the Lancashire Miners' Vote (1890-1906)

The 1890s witnessed the political birth of organized labour, a sustained and enduring interest in parliamentary affairs and a desire to participate in them. The miners were to the forefront of this general movement. They had a tradition of political activity. Much of their efforts in the 1840s had been concentrated on attempting to obtain improved mining legislation. At the 1874 general election, when the first workingmen were elected to the House of Commons, both happened to be miners. By 1886 six of the ten M.P.s who were workers came from the coal industry. The degree of political involvement displayed by miners arose from the large number—indeed, the growing number—of state regulations which affected their working lives. Their fate was bound up with the twists and turns of the parliamentary pen.

Miners were favourably situated to have a political impact. Compared with most other trade unions, their organization was strong numerically and financially. The membership, mostly living in clusters of mining villages, formed a big proportion of the electorate in a number of constituencies. These were potentially winnable seats. Further parliamentary representations appeared to be highly probable in the future.

But, while this seeemed to be the national picture, the position in the North-West was rather different. Lancashire miners felt the need for political involvement more than their compatriots. The weakness of the union *vis-à-vis* the employers made them hanker after legislation all the more. Certain troubles appeared only curable with medicine dispensed through Act of Parliament. These included:

(1) The large number of men killed or maimed in the mine. Lancashire's accident rate was forty per cent above that of Northumberland.[1] Only more stringent regulations, strictly enforced, would compel the small coalowners of North Lancashire to spend money, transforming their pits from death-traps into safe, well-ventilated establishments.

(2) The Eight-Hour Day. Hours of labour were longer in Lancashire than any other English coalfield.[2] Obdurate owners seemed determined to keep it that way. The union had little prospect of securing an appreciable reduction of working hours; parliamentary action appeared to be the only hope.

(3) Employers' liability for industrial injury. After the passing of the 1880 Act, many Lancashire coalowners compelled their men to contract out and join the Lancashire & Cheshire Miners' Permanent Relief Society as a condition of employment. Thus, a lot of colliers were denied the provisions of the Act and corralled in a Society which, to many miners, seemed to benefit the bosses more than themselves.

(4) The appointment of checkweighmen. At many Lancashire collieries these officials still did not exist in the 1890s. Employers used loopholes in the law to prevent their appointment.[3] But even where they did exist, the management frequently influenced the election of checkweighmen with threats, bribery and dismissals, or, alternatively, refused to provide facilities for the election to take place.[4]

All these four issues tended to make miners' minds dwell on things parliamentary. But, if the need for political activity was greater in Lancashire than elsewhere, then equally the obstacles to it were also greater. Union leaders were faced with problems that did not exist in other coalfields.

The first arose from the fact that miners tended to be scattered throughout the county. Only in two constituencies, Leigh and Wigan, did they constitute the majority of the electorate.[5] Traditional mining seats, as known for instance in South Wales, were almost entirely absent from Lancashire: there were no constituencies where the union could dictate who became Members of Parliament. To complicate the situation further, there were few mining villages in the North-West. Most miners lived in communities where they intermingled with the rest of the county's population. As a result, there was not the same degree of group solidarity and collective conduct as characterized other coalfields, where people lived together, worked together, and then voted together.

In other coalfields, political evolution followed a similar pattern. Once they gained the vote, colliers expressed their anti-Tory feelings by voting for Liberal candidates. It became, therefore, relatively easy for the union to pursuade them that the candidate upon whom they should bestow their favours should also be, at the same time, a miner. In most parts of the country, the acceptance of Lib.-Lab. candidates—

men like Burt and MacDonald—involved little or no internal dissension.

There could be, however, no easy road for the Lancashire Federation. It was confronted with a unique phenomenon—in contradistinction to elsewhere, the Lancashire coalfield had many pitmen who were Tories. Even union officials who disliked it had to acknowledge this fact. When, in 1895, a miners' delegation met Lord Salisbury, Thomas Aspinwall told him: 'I come from Lancashire ... I can verify that you have a large number of supporters in Lancashire ... I know that you have thousands in connection with the mining industry in Lancashire who are very ardent supporters of yours.'[6] Aspinwall had good reason for knowing the truth of his own statement: he had just lost the mining seat of Wigan to a Tory in the 1895 general election.

A similar situation existed elsewhere in Lancashire. The mining constituencies continued to return Conservative M.P.s Even solidly proletarian places like St Helens remained Tory until 1906. This was partly because of the skilful way the Conservatives exploited the Irish immigration issue but, probably more important, it was due to the records of successive Conservative governments.[7] Legislation on trade unions, factory conditions, mining safety, etc., all had a powerful effect. Many Lancashire miners agreed with the remark of Alexander MacDonald, 'The Conservative Party have done more for the working classes in five years than the Liberals have in fifty.'[8] Even opponents of the Conservatives, like the journal, the *Workmen's Times* had to admit the Tories had the image of being a reformist party:

> 'The Conservatism of the Lancashire working man has not its roots in any deep and abiding affection for the Throne, the House of Lords, or the Established Church. Its main cause is to be found in a remnant of gratitude for the passing of the Factory Acts against the strenuous opposition of the employing class, mostly Liberal in politics.'[9]

The textile manufacturers' close connections with the Liberal Party provided the Tories with a bountiful supply of political ammunition. All the tensions generated in the mills, the operatives hatred for the Gradgrinds, were exploited by the Tories in their attacks on the Liberal Party. They could use the language of working class protest. Even when leaflets were not officially produced by the Conservative Party, they were not averse to providing the financial resources for the discontented to air their grievances against their employer—so long as he happened to be a Liberal parliamentary candidate. The technique is exemplified by the handbills reproduced on the opposite page. They show not merely how the anger of miners was kindled against John Lancaster, a coalowner, who contested Wigan four times for the Liberals, but also how John Lancaster, by his anti-working class activities, gave his political enemies such magnificent material.

Many Liberals acted in this way. As a result, they gave Conservatives considerable opportunities, which, as Dr John Vincent points out, were effectively used:

> 'The Lancashire Conservatives' success in adapting to industrial conditions was in many cases so closely tied, whether by coincidence or otherwise, to their development of a populist political language, the cultivation of the common touch, and their tactical alliance with aspects of the working class movement.'[10]

In addition, Tory popularity thrived because of the adept way they handled the immigration issue. By 1880, more than a third of Irish people living in England were to be found in Lancashire, making it 'the most Catholic county in England'.[11] The indigenous population expressed its hostility in terms similar to those employed against coloured immigrants today: the Irish, it was claimed, were dirty, had large families, and had a different way of life. Worst of all, they increased the supply of labour and thus helped to depress wages.[12] Professor Pelling argues that St Helens, an overwhelmingly working-class constituency, nevertheless remained 'a marginal Conservative seat due to religious conflict—itself a result of the discords caused by heavy immigration'.[13] The same applied to Wigan, equally working class as St Helens and with the same proportion of Irish: Conservative M.P.s were repeatedly returned and even survived the great Liberal landslide victory of 1906. Where the percentage of immigrants was high, the Conservative vote was also high. Table 2 indirectly confirms this fact. Those constituencies with large Catholic populations—and therefore, presumably, in most instances with a lot of Irish—were generally those seats where the Liberals did badly.

Obviously, the Lancashire Miners' Federation had to proceed into the political arena with extreme caution. An agreement with the Liberal Party, the running of Lib.-Lab. candidates, was sure to arouse the ire of masses of Tory miners. But besides its exposed right flank, the Federation found its left flank attacked by socialists, out to denounce any shabby compromises with the Liberal Party involving the sacrifice of principles.

The threat from the left had to be taken seriously. Socialist ideas acquired considerable support amongst the miners. This was particularly true of the young, articulate colliers. They were the kind of men who led the struggle for union recognition against large coalowners like Andrew Knowles and fought against the persecution of the masters in the Burnley area. Once a union lodge was established, they usually played a prominent part, becoming the branch officials. So they tended to gain publicity and influence out of all proportion to their numbers.

The campaign for socialism had begun in Lancashire in the 1880s. The formation of the Social Democratic Federation almost coincided with the massive strike of Lancashire miners in 1881, objecting to their employers compelling them to contract out of the Employers' Liability Act. H. M. Hyndman took the miners' side in the columns of the *Pall Mall Gazette*, maintaining that the strike was 'in the interests of all England'.[14] Hyndman also wrote to Karl Marx:

> 'Matters are ripening in the North of England. This weather will push them yet faster. Poor creatures with the pits closed and their children crying for food and warmth! ... We shall be into revolution before we know it. That indeed is coming, but I want it to come peacefully. I am going to see some of the leaders of the working classes ... The movement has begun for which you have so long waited: now it has to be guided to grow.'[15]

In accordance with this policy, Hyndman's paper, *Justice*, paid particular attention to mining matters. Eventually, it gained a response in the form of socialist activities. On 28 March, 1885, *Justice* reported that 'trade is very bad' in St Helens, with miners on a four-day week; S.D.F. literature was being read with interest. Then came accounts of public meetings at Eccles, Pendleton and Oldham. At Pendlebury a branch was formed. The secretary reported, 'Many more would join only they are afraid of the local magnates, the colliery proprietors.'[16] Two years later, John E. ('Jack') Williams, who was a national figure in the S.D.F., visited the Pendlebury branch. He stated that its thirty-two members were miners and cotton spinners; pitmen in the district were poorly paid and had no union.[17]

Persistent campaigning began to achieve results. Dona Torr wrote, 'South Lancashire early became and always remained the strongest provincial centre of the S.D.F.'[18] This can partly be attributed to the activities of Tom Mann. He spent most of the year 1888 living in Bolton.[19] A fine orator, he regularly addressed large gatherings in

Bolton and nearby towns. The May Day rally was most impressive. *Justice* reported that a crowd estimated to be twenty thousand walked to Blackedge, a place on the Lancashire-Yorkshire border that used to be the site for Chartist demonstrations.[20]

But it was far from being all the work of Tom Mann; many others helped the socialist movement to progress. Amongst them were a number of talented miners. The first Lancashire organizer of the S.D.F. was M. K. Hall. He gained the position, said the *Workmen's Times*, because of:

> 'his knowledge of economics and by his uncompromising advocacy of the principles of Social Democracy. While in receipt of the low wage of eighteen shilings per week as a miner, he was frugal enough to save sufficient to leave his work and devote himself to study for twelve months. He gave his attention to classics and economics. In appearance Hall is rather singular. You would hardly imagine that he was a student; but you could easily believe he was a navvy or a bricklayer's labourer. His face is pinched and drawn as if he has known hard times.'[21]

Then there was Albert Ward, a collier from Bryn, near Wigan. He stood for the local council against the managing director of the colliery where he worked. He failed to gain the seat, but he did secure the sack! Subsequently, he developed into a well-known open-air speaker of the S.D.F., 'one of their best propagandists' says Walter Kendall.[22]

The vital thing was that the S.D.F. learnt to speak to the miners in their own language. This was especially valuable during the 1893 lock-out. By that time, the S.D.F. had twenty-one branches in Lancashire; the newly-formed Independent Labour Party had eight.[23] The existence of two socialist organizations did not create discord, diminishing the power of the socialist appeal. Instead, they tended to complement each other, the I.L.P. gaining members where the S.D.F. could not. Although the national leaders quarrelled, there are very few examples of it being transferred to local level. As the *Labour Leader* reported, 'The S.D.F. and I.L.P. in Lancashire are working together splendidly.'[24]

This unity in action was displayed during the 1893 lock-out. Both socialist groups had miners who played prominent parts within their union. Those who were not connected with the coal industry still endeavoured to identify themselves with the miners' struggle, holding meetings to show solidarity, collecting money for the union's fund, and helping with relief work. In Manchester the police arrested socialists for making street collections. But this proved to be no deterrent. 'If the strike continues,' warned the *Workmen's Times*, 'the Manchester Watch Committee will be forced either to withdraw their prohibition or they will have to put the whole of the members of Manchester and Salford I.L.P. in prison.'[25]

The dedication of the socialists, the persecution of them by the authorities, their allegiance to the union's cause—all these things helped to endear them to large numbers of miners. Once the lock-out ended, socialists began reaping their reward in electoral terms. In October 1893, Jesse Butler, a miners' agent, stood as the I.L.P. candidate at Openshaw. The *Labour Leader* reported that he was 'stirring up the electors with a long pole'.[26] After a vigorous campaign, he secured a seat on Manchester City Council. In November 1893, another miner, John Sparling, was returned as the S.D.F. candidate to Burnley Borough Council in a ward where he previously had been defeated. 'Thanks to the miners' lock-out for such a result', wrote the local correspondent to the *Workmen's Times*. Comrade A. G. Wolfe permeated socialism among the miners. 'They not only voted for us, but worked with enthusiasm and determination never excelled.'[27]

Perhaps it has not been fully appreciated the extent to which, in that period, the British socialist movement was a regional phenomenon, almost confined to a few parts of the country. This is shown by an examination of the I.L.P. membership returns: during 1893 and 1894 *The Clarion* published each week the names and addresses of the people joining. It reveals that Lancashire and Yorkshire were the I.L.P.'s two great strongholds, with roughly a quarter of the entire membership in each. The general election of 1895 confirmed this predominance: of the 43,498 votes cast for socialist candidates, 26,605 were won in Lancashire and Yorkshire.[28]

The position in the S.D.F. was even more remarkable. This can be shown by juxtaposing two quotations from Dr Tsuzuki's book on *H. M. Hyndman and British Socialism*: 'In October 1893, the *Burnley Socialist and North-Eastern Lancashire Labour News* was founded as the organ of six local branches of the S.D.F., whose aggregate membership had now reached nearly two thousand.' Later he cites the S.D.F.'s 1894 conference report statement that total membership was four thousand.[29] In other words, we find the hardly credible situation that nearly half the S.D.F.'s membership belonged to six branches in North-East Lancashire. When it is remembered that the S.D.F. was also reasonably strong in other parts of the county, it can be seen what a stronghold of the socialist movement Lancashire really was.

An editorial in the *Workmen's Times* was headlined 'Will Lancashire lead the way?' It referred to the vigorous socialist agitation taking place. Branches of the Fabian Society, I.L.P., and S.D.F. 'have sprung up on every hand, and even trade unions and Trades Councils are crying out for Independent Labour representation'.[30]

It was against this background that the Lancashire Miners' Federation had to take its decision about political activity. The presence of large numbers of Conservatives and socialists in its ranks constituted a highly combustible mixture, particularly as the plan was to have an alliance with the Liberals. The union did not even possess a unified leadership, sufficiently strong to be able to weather the inevitable storm of criticism and give a sense of direction to the policy. The disagreements among the rank-and-file also found expression among full-time officials:

(1) Lib.-Lab.	Sam Woods, Thomas Aspinwall, Robert Isherwood.
(2) Socialists	Jesse Butler, Tom Greenall.
(3) Conservative	Thomas Ashton.

As Federation secretary, Thomas Ashton was favourably placed. Outspoken opposition from him could have made it extremely difficult for advocates of the Lib.-Lab. policy. But Ashton, a man with intense loyalty to the union, wanted to avoid having bitter clashes with fellow officials. He contented himself by making an occasional oblique comment, advising the union to be cautious about its political involvement. Sometimes, however, Tory undertones crept into his secretary's reports. For instance, in 1897, when the Eight-Hour Bill was defeated on its second reading in the House of Commons, Ashton suggested that miners should not be disheartened by the news since '*under a Conservative Government* the majority had been reduced down to 41'.[31] (My emphasis RC.)

But even had Thomas Ashton desired to prevent the Lancashire miners running parliamentary candidates, he would have been unable to do so. The union's de-centralized structure put the power in the hands of the districts, which could initiate schemes without informing the county Federation.

The first steps towards entering the parliamentary arena came from the biggest district union. On 29 March, 1890, Ashton and Haydock held a long discussion on 'the necessity of direct Labour Representation in the House of Commons'.[32] Delegates decided to go ahead with the venture, irrespective of the attitude of the Lancashire Federation. Sam Woods was selected as the candidate. All lodges within the district were asked to approve the decision and to make local soundings, so the most promising constituency could be selected. A month later, Ashton and Haydock decided to contest Ince. A specially elected parliamentary committee, with Stephen Walsh as secretary, resolved to create an election fund. Each member of the district was to be levied a penny a month. It was only at that meeting, held on 1 May, 1890, that it was decided officially to inform the Lancashire Federation.[33]

When the news was brought before the Federation conference, there was a confused and heated debate. Ashton and Haydock had asked it to give moral and financial support. The big question was: Should the Federation do more or less than that? Standish delegates wanted the Federation to negotiate with Ashton and Haydock to take over responsibility for conducting the campaign. It would mean that the full financial resources of all the district unions could be tapped. At this point, St Helens and Hindley representatives made it plain they did not want to pay any political levy, but they were told they would be disaffiliated if they did not. Thomas Aspinwall, who brought to the meeting a £15 donation from the Liberal Party as encouragement, told delegates that a lot could be gained through political activity. The majority agreed with him.[34]

The fact that the Federation took over from Ashton and Haydock does not appear to have had any effect on Wood's campaign, which began immediately although it turned out to be two years before election. Sam Woods issued a statement, giving his reasons for standing. He pointed out that, while other classes were well-represented in Parliament, 'the labour interest is practically without representation'. In Lancashire there was not a single working-man M.P. The Member for Ince, a local coalowner, never voted for progressive legislation brought before the Commons.[35]

Soon the campaign ran into difficulties. To have a chance of winning, Woods had to secure the votes of the majority of miners who, in previous elections, had voted Tory. He could not afford to do anything that might antagonize them. As the *Labour Tribune* observed, 'Several members of Mr. Woods Election Committee are Conservative Working-men. And if Mr. Woods is returned for Ince, it will be by labour votes, regardless of party.'[36] With this in mind, it was thought advisable to play safe: 'In the opinion of the Committee, which was elected pro tem from the members of his own district until the Federation took the matter in hand, it was thought the wisest course to remain neutral on most contentious subjects.'[37]

Some controversial issues, however, could not be side-stepped. With so many Irish immigrants residing in the constituency, he had to take a line on the question of Home Rule. To oppose Home Rule would alienate traditional Liberal supporters, as well as the Irish themselves; to support it (as he eventually did) would make it just that much harder to win over the Conservative miners.

Sam Woods' stand on Home Rule made it even more imperative for him to assuage the fears of Tory colliers that he possessed no extreme radical tendencies. Likewise, he had to convince middle-class Liberals hesitant about backing a miner, that he accepted the principles of bourgeois respectability as firmly as they did themselves. So Woods, who was sincerely anti-socialist, discovered it was tactically advantageous to emphasize this point. Socialism, he declared, was cemented with violence and attacks on property. This remark naturally aroused the anger of many socialists, both locally and nationally. In a tart letter to the *Labour Tribune*, Cunningham Graham pointed out that 'Woods says, with great humility, "it will be an enormous advantage if I am returned to Ince", but he does not say for whom the advantage is to be, whether for himself or the electors.'[38]

In spite of Woods' anti-socialist attitude, the prevailing opinion in left wing circles was one of tolerance. Allowance was made for the awkward position he was placed in at Ince by the Tory miners. Also, it was thought that, whatever Woods' limitations, his election would be a step forward, an improvement on the present incumbent. So Keir Hardie sent greetings to him, expressing the hope that he would win the seat and help 'to form one of the new Labour Party, which will gather round Cunningham Graham and Ben Pickard in the next Parliament'.[39]

There was a second chance of labour representation coming from Lancashire as a result of the 1892 general election—Thomas Aspinwall was fighting Wigan. The agreement between the Liberal Party and the miners' union was reached much later than that over Woods' candidature. It may be that he appealed to the Wigan party bosses after the Liberal success in the Rossendale by-election. This was attributed mainly to the colliers' vote. Perhaps they thought, with a popular miners' agent like Aspinwall, the same thing could be done in Wigan. Or, conceivably, the Liberals might have considered it a good idea to have a candidate similar to Woods since the nature of the two constituencies were remarkably alike. In both the underlying enemy was the same—the Wigan Coal & Iron Company. 'This overgrown and bloated corporation,' declared the *Wigan Comet*, 'claims for itself the right of returning three Members of Parliament, namely those nominally returned for Wigan, Ince and Westhoughton.'[40]

Aspinwall's opponent, the Honourable Member for Wigan, was Francis Sharp Powell. He fought the seat after the sitting Member, Lord Lindsay, chairman of the Wigan Coal & Iron Company, had been elevated to the House of Lords. By accident, the by-election of January 1881, coincided with the bitter strike of colliers. As a result, there was a spectacle probably unique in British history: Tory and Liberal coalowners competing with one another to wine and dine their striking miners. Powell's generosity appears to have exceeded that of his opponent, John Lancaster. But, alas, it did not receive its true reward. After a nine-day court hearing, he was unseated. In giving judgment, Justice Grove said, 'I can certainly say this, that there is reason to believe corrupt practices did extensively prevail.'

Reporting the verdict, the *Manchester Courier* commented: 'Not a little sympathy was expressed for Mr. Powell, who at enormous cost has fought several great electoral battles. His outlay in these contests is estimated at between £50,000 and £60,000.'[41] But only a small proportion of this sum appears to have been spent at Wigan. As the *Manchester Examiner* pointed out:

> 'Wigan is small beer by the side of Oxford, Macclesfield and Gloucester. There was no sumptuousness of corruption in Wigan. The mass of the voters were very poor and pettifogging bribes sufficed. But, taking breakfasting, treating and special transactions together, it was quite large enough to account for Mr. Powell's majority.'[42]

Subsequently, Mr Powell learnt how to control his generosity, dispensing it between elections rather than on polling day. In 1885, he was returned for Wigan, where he became extremely popular.[43] In part, this occurred because he continued to bestow gifts—he gave the town a reading room—but also it arose from his enlightened approach. He acquired the reputation of a reformer. In Parliament, he introduced the Public Health Amendment Act of 1890, worked on the Cotton Cloth Factories Act of 1889, and was an enthusiastic supporter of all educational improvements. So, when the general election of 1892 came, he could rely upon his own personal political standing. There was no bribery. The only irregularity appears to have been a little bit of discreet intimidation by Wigan Coal & Iron Company. The *Wigan Comet* complained: 'The presence about almost every polling station during the day of the election of a crowd of the officials of this Company is, in itself, an insult to every independent voter.'[44]

The results of the two contests were:

Ince		*Wigan*	
Sam Woods (L)	4,579	E. S. Powell (C)	3,422
Col. Blundell (C)	4,352	T. Aspinwall (L)	3,312
Majority	277	Majority	110

The election campaign had taken place amid rising dissension in the union's ranks. Once it was over, and Woods had defeated his Conservative opponent, criticism became more vehement. The editor of the *Wigan Comet* reported: 'I hear with regret that a considerable number of miners are leaving and have left the union since the election. They are Conservatives, of course.'[45] He went on to say: 'Several new miners' clubs have already been formed in Wigan and district, and there can be no doubt the movement will spread.' His forecast proved correct. On 12 November, 1892, these clubs combined to form the Conservative Labour Union.

The 'blue button union', as it was known, aimed to recruit textile workers and others as well as disgruntled colliers. One of its leaders was Tom Oakes, of Hindley, one of the militants who led the strike of 1881. Trade unions, the new organization declared, should be non-political. The Lancashire Miners' Federation had been wrong to spend £1,000 on the election expenses of Aspinwall and Woods—'squandered to gratify the vanity and ambition of a few self-seeking individuals'.[46] It had, therefore, to be taught a lesson, recognize the error of its ways, and revert to its customary policy of being non-political. The Lancashire Miners' Federation would only do this if Sam Woods was soundly defeated at the next election. So the Conservative Labour Union saw nothing contradictory in stating that it was a non-political body while, at the same time, declaring its primary aim was to see Woods defeated at the next election.

For a while, the Conservative Labour Union appeared to be a serious threat to the Lancashire Miners' Federation. Hostile journals acknowledged this fact. In January 1893, the *Wigan Observer* reported Oakes addressed one of its meetings in the town, which was well-attended.[47] In March 1893, the *Wigan Comet* said several 'discarded compositors are making a success of the Conservative Labour Union'.[48] More important, the 'blue button union' became of sufficient consequence to draw the fire of the miners' union nationally. It was attacked at the 1893 Birmingham conference and the *Labour Tribune*, which acted as the national organ of the mineworkers, denounced it: The Conservative Labour Union, it said, was 'a sham union, championed purely in the interests of the employers'.[49]

Besides engaging in debate, the miners' union encouraged its members to take action against the Conservative Labour Union. For instance, when a man at Colonel Blundell's Pemberton colliery joined the 'blue button union', the rest of the labour force issued an ultimatum. They resolved 'to stop the colliery unless the employer will do one of two things, namely, stop the man or force him into the organisation proper'.[50]

Under attacks such as this, the Conservative Labour Union seems to have collapsed. One of the reasons might well have been that Tory employers saw its existence created disharmony, industrial discord which was out of all proportion to the political gains made. Indeed, the political influence of the 'blue button union' appears to have been negligible. It did not dissuade the miners from running parliamentary elections nor does it seem to have materially affected the election results. Furthermore, there was always the danger that, by delving in trade union affairs, the Tories might receive a painful shock. If the Conservative Labour Union were to exercise the traditional function of a trade union—coming out on strike—then the Tories might find themselves financing a struggle against themselves!

But the Lancashire Miners' Federation did not merely have to worry about the 'blue button union'. Politics inspired industrial schisms at the other end of the spectrum. In February 1892, W. T. Postlewaite, a Fabian schoolmaster, tried to persuade the unorganized colliers of Pendlebury to join the Independent Labour Union.[51] This was an organization which advocated nationalization of the land, mines and railways, and aimed to get workingmen elected as councillors and M.P.s. The Swinton & Pendlebury branch, which recruited a hundred members in the first month of activity, was number nine branch of the Independent Labour Union—so, presumably, somewhere there was at least eight others. In April 1892, Postlewaite, who was described by the *Workmen's Times* as 'one of the ablest speakers in the North of England', won a county council by-election. He had a majority of 747 out of a total vote of 2,861 in a 75 per cent poll.[52]

It may well be that another union had, at least partly, an element of political protest in its motivation. The Leigh Reform Association, a breakaway which was principally a reaction to the Lancashire Federations' reorganization plans (see Chapter thirteen), had Robert Grundy as one of its two prominent leaders. He was an S.D.F. member, who unsuccessfully stood for Leigh Borough Council.[53]

While all these attempts to form rival unions must have caused disquiet and concern to the Lancashire Miners' Federation, none of them gained sufficient strength to become viable. Their numerical weakness naturally found reflection in financial weakness. As a consequence, they could not provide the same protection for their members as the Miners' Federation. Their programmes, moreover, always contained a divisive element, limiting their appeal. It precluded them winning over vast numbers of workers. Most Lancashire colliers regarded them as a source of disunity and hence were not prepared to join.

Nevertheless, the general climate of opinion produced by these rival bodies made things very difficult for the Miners' Federation. For every man who joined, say, the 'blue button union', there must have been many more Conservative colliers who remained aloof, joining neither it nor the Miners' Federation. In a handbill, printed on 12 April, 1893, Thomas Ashton conceded this point. It began by admitting that large numbers of miners in Lancashire were not in the union and excused themselves on grounds of politics. He attempted to justify the Federation's Lib.-Labism in the following terms:

> 'We have members of the Federation of all shades of political opinions and no man's conscience is interfered with. Certainly we paid the election expenses of Mr. Woods, who received the support of the Liberal Party in Ince, but it is well known that up to the present it has been utterly impossible to return a "Labour Candidate" without the assistance of one of the great political parties.
>
> 'Mr. Woods is a Labour Liberal member—Labour first—and he has done more for Labour during his time in Parliament than any other member of the House of Commons.'[54]

Whether or not Conservative colliers outside the miners' union were convinced by Thomas Ashton's explanation, many non-Conservative colliers inside the union were not. There was a grave threat to the Federation's leadership from members who, in a crude way, combined militancy with Marxism. As shown in Chapter thirteen, one of the main spokesmen of this tendency—John Sparling—was nominated by the largest district union for the office of Federation secretary. Thomas Ashton beat his challenger, but nevertheless, must have been disconcerted with the strength of the opposition. These left-wing critics were at variance with Federation leadership politically as well as industrially. Just as they thought that the failure to fight the large coalowners, like Andrew Knowles, or to escalate the struggle against the owners of North-East Lancashire were signs that the Federation leaders were too prone to class collaboration, so they were prepared to subject the alliance with the Liberal Party to scathing scrutiny.

In face of this onslaught, a Lib.-Lab. M.P. like Sam Woods was highly vulnerable. He had to serve two masters—the Liberal Party and the Miners' Federation. The one provided him with the political machine that put him in Parliament; the other gave him the votes and finance. Consequently, he had to remain on good terms with both organizations. But when they quarrelled, Sam Woods' position became hardly tenable.

This is precisely what happened during the miners' lock-out of 1893. To quell disorders at Featherstone, Yorkshire, the troops fired on the crowd, killing two miners and wounding sixteen. Since a Liberal government was in office at the time, it was held to be responsible. Mr Page Arnot writes:

> 'The shooting at Featherstone caused widespread anger amongst the miners. Considerable anxiety was expressed in Parliament because troops had been used in connection with the mining lock-out. It was said that not for fifty years in Great Britain had troops fired upon the populace; and memories were revived of the repression in Chartist times.'[55]

In the miners' opinion, the main culprit was H. H. Asquith, the home secretary:

> 'An effigy of Asquith, surmounted by a Death's Head and cross-bones, was carried to Trafalgar Square in London. For long afterwards Asquith's name in the minds of many workers was associated primarily with the shooting down of unarmed miners.'[56]

Sam Woods, M.P., had a much more restrained response to the Featherstone massacre. In a letter to Asquith on 21 September, 1893, he merely expressed regret at the unfortunate occurrence and concluded by saying:

> 'While writing you on this matter, allow me to express, on behalf of the Lancashire miners and their leaders, their sincere thanks for the flattering opinion you expressed as to the peaceful conduct of the miners during the present unfortunate lock-out.'[57]

The editor of the *Labour Elector* sarcastically commented:

> 'Mr. Samuel Woods, M.P., is really too good for this world. His love and admiration for his (Liberal) enemies and for the enemies of his class are altogether too heavenly and quite out of place in a mere earthly miners' leader.'[58]

Likewise, the *Workmen's Times* attacked Woods for not severing his Liberal Party connections after Featherstone.[58]

Again, in 1895, he came in for severe censure. When a motion opposing the introduction of the Eight Hour Day came before the House of Commons, both Woods and Pickard voted with the majority; only Keir Hardie voted 'No'. This provoked the *Labour Leader* to exclaim: 'When the miners return I.L.P.ers to represent them, they won't have to go cap in hand. Until they are prepared to do that, they had better not have the name of being represented.'[59]

Sam Woods' entanglements with the Liberal Party sometimes led him to support candidates hostile to many of the aims of organizations to which he belonged. This happened, for instance, at Darlington in 1898. It provoked a local Boot & Shoe union to talk of 'his seeming indifference to carrying out the resolutions of the Trades Union Congress ... He has forfeited the confidence of the Trade Unionists of Great Britain.'[60]

Sam Woods' conduct led to widespread criticism. Keir Hardie described him as 'a hooded falcon guaranteed to return to rest when called upon by his keeper'.[61] Beatrice Webb found him 'intensely conservative and difficult to win even to step by step reform'.[62] The *Wigan Comet* reported a discussion among miners in which he was referred to in disparaging terms. One collier remarked, 'You must admit that Messrs. Woods and Aspinwall seemed very small and insignificantly indeed beside Tom Mann ... After hearing Mann, I must admit I was heartily ashamed of our own local leaders.'[63]

Obviously, such criticism was damaging to an M.P. like Sam Woods. But the bankruptcy of the Lib.-Lab. approach was proved not by its opponents so much as by the Lib.-Lab. M.P.s themselves. They did not secure the parliamentary reforms. They did not secure sufficient parliamentary seats. In other words, the Lib.-Lab. strategy was doomed by its own failure.

The case of Thomas Aspinwall illustrates this point. He was a model working-class Liberal candidate—a well-known miners' agent, with orthodox Liberal views, whose union paid the election expenses.

After he reduced the Conservative majority to 110 votes at the 1892 general election, the local Liberal paper proclaimed, 'We shall hear less in future of Wigan being put down as a safe Tory seat.'[64] The Liberal politician, H. H. Asquith, thought the result highly commendable. He told a public meeting in Wigan that 'with better organisation, victory could have been achieved'.[65]

In these circumstances, it might be thought Aspinwall was the obvious choice for the next election. But this was not so. The local Liberal Association wanted W. Woods, a coalowner and former Mayor of Wigan, as the candidate. This led to the intervention of the Liberal Party nationally. Lord Rosebery stated that he thought Aspinwall should contest Wigan again. So Woods withdrew his candidature. But the Wigan Liberals, many of whom had never been reconciled to having a mere miner to represent their cause, still refused to support Aspinwall. They passed a resolution to this effect by thirty-nine votes to nine despite the fact that they had no alternative candidate.[66]

The decision of the Wigan Liberals stimulated much comment. The *Weekly Times* saw it as an example of middle-class prejudice which would not fool the workers. The *Sunday Chronicle* declared that the Wigan Liberals had not as much discretion as a Chinese general and asked, 'How long will it be before the Liberal Party comes to a great and glorious smash?'[67] Liberal peer, Lord Ripon, also saw this danger. The decision was 'a very great mistake'; unless the Liberals were prepared to adopt candidates like Aspinwall, the working class would turn to the I.L.P.

It is true that the animosity of the Wigan Liberals pushed the Lancashire miners along the road to independent political activity. In January 1895, the West Lancashire district decided not to allow Aspinwall to withdraw his candidature.[68] Even if the Liberal Party did field a candidate of its own, Aspinwall was still to fight the seat. The Federation endorsed this move.[69] So did the Wigan Trades Council. His campaign was run by working-class organization. The entire Labour Movement of the district supported Aspinwall, except for the I.L.P. and S.D.F., who counselled their followers to abstain in Ince and Wigan.

At the general election, July 1895, the results were:

Ince		*Wigan*	
Col. Blundell (C)	5,235	F. S. Powell (C)	3,949
Sam Woods (L)	4,790	T. Aspinwall (L)	3,075
Majority	445	Majority	874

For both the Liberals and the miners, these results heralded the end of the misalliance, which could never be described as a love-match or a wedding of like minds. The Liberals could argue, on the basis of the Ince result, that even where there was harmony between the miners' union and the local Liberal Association it was still futile. The union could not control its members electoral habits, preventing large numbers from voting Tory. The Lancashire Miners' Federation, on the other hand, could point to the lesson of the Wigan contest: given the choice, many middle-class Liberals would rather see a Tory retrned than a worker with Liberal views. The same attitude, of course, was also present elsewhere. So even if it were possible to stand on a Liberal ticket, as at Ince, the advantages to be derived from such a move appeared to be negligible. Never again did the miners strive to reach agreement with the Liberals in either constituency.

Table 17

The Liberal vote in South & Mid-Lancashire, 1885-1906

Percentage of Catholics in the constituencies	Percentage Liberal vote
Over 20%	40·6%
10—20%	45·8%
0—10%	50·1%

Table 18 *The Liberal vote in Lancashire, 1885-1906*

Cotton Region		*Mining Region*	
% of electorate textile workers	Liberal vote	% of electorate miners	Liberal vote
+30%	51·8%	+40%	47·6%
20—30%	47·3%	25—40%	47·0%
10—20%	52·1%	10—25%	46·3%

I am indebted to Mr Jeff Hill, a Ph.D. student at Keele University, for the statistics contained in Tables 18 and 19.

CHAPTER FOURTEEN

From Liberalism to Labourism

The repudiation of the Liberal Party, itself a negative move, did not furnish the basis for a new policy. It left the Lancashire Federation in limbo, uncertain what to do. The lack of a clear, obvious course of action strengthened the hands of the hesitant. Those still wedded to the idea of a Lib.-Lab. alliance strived strenuously to re-establish the broken links. But in vain: after 1895 no miner stood for Parliament under Liberal colours in Lancashire.

In most other parts of the country the electoral alliance remained viable, and a leader like Sam Woods had a personal interest in its perpetuation. In 1897, he was the victorious candidate of the Lib.-Labs inside the trade union movement for the post of secretary of the Parliamentary Committee of the T.U.C., narrowly defeating his socialist opponent, Tom Mann. By attaining national prominence, Sam Woods acquired contacts elsewhere, providing him with information of selection meetings where no other Lancashire miner stood the remotest chance of nomination. Thus, it was that he became the Lib.-Lab. candidate at the 1897 by-election for the London constituency of Walthamstow. He won the seat, but was subsequently defeated at the general election of 1900.

After that Sam Woods, despite his influential friends, was unable to find a constituency. Campbell-Bannerman and the national Liberal Party organization wanted him selected for the Dewsbury by-election of 1902.[1] Tom Greenall, in the *Pendlebury Pioneer*, described how Sam Woods journeyed over to Yorkshire, met the selection committee, 'and at their invitation spoke for about two hours, explaining his political opinions from the days of Adam down to the present. They were most friendly and sociable, and he came away with the impression that he would receive a call to be the candidate.'[2] But Dewsbury Liberals preferred Walter Runciman, the son of a wealthy shipowner, to Sam Woods. The *Pendlebury Pioneer* suggested this was because of middle-class prejudice. Another possible explanation is that the selection meeting retaliated for being subjected to a two-hour ordeal!

Nevertheless, there can be no doubt that there was a pervasive feeling of revulsion on the part of Liberals at constituency level to selecting miners as their candidates. This 'stupid and grudging attitude of local Liberal associations', as one historian described it, made it increasingly difficult for the various county unions to retain their Liberal alignment.[3] What the Lancashire miners discovered in the 1890s, miners elsewhere discovered roughly ten years later. As a result of the hostility manifest by the Liberals, the L.M.F. was always a front runner in the race towards independent politics.

In 1900 the Lancashire Federation resolved to run its own candidates. For these to be successful, political allies had to be sought. So the L.M.F. sent delegates to the inaugural conference of the Labour Representation Committee. It was the only miners' union to have delegates there. Joe Greenall topped the vote out of 22 candidates for trade union representation on the L.R.C. national committee.[4] But he had to resign for the Lancashire Federation, having second thoughts, decided not to affiliate. To be anti-Liberal is not the same as being pro-Labour.

The question, however, could not be left there. In May 1903, the Lancashire Federation decided to adopt four candidates, and this obviously raised the issue of its relations with other organizations running working-class candidates. Consequently, it was resolved to re-join the Labour Representation Committee. But it may be significant that, while the Lancashire Federation affiliated to the full extent of its thirty-seven thousand membership, it sent a low-powered delegation to L.R.C. next conference: Ashton, Aspinwall and Woods did not attend.[5]

The Lancashire Federation created problems for the Labour Representation Committee, although doubtless, on balance, its affiliation was greatly appreciated. The difficulties arose over tactics. The L.R.C. took the position that, while it was vital to preserve political independence, this did not mean it would have to remain isolated. Hence, the L.R.C. was prepared to enter into electoral agreements with the Liberal Party. These, it was thought, would be mutually advantageous since it would stop the anti-Tory vote from being split.[6] But, as we have seen, this was far from being the attitude of the Lancashire miners. They had no compunction whatsoever about fighting Liberals. They regarded them as bad, perhaps worse, than the Conservatives.

The conflict of approach became clear in 1903, when the Lancashire Miners' Federation decided Tom Greenall should stand at Accrington. The L.R.C. wanted the seat left for a Liberal candidate to contest. Ramsay MacDonald wrote to Greenall, suggesting he retire from the contest at Accrington and fight South Salford instead.[7] Yet, despite being subjected to this pressure, Greenall remained adamant. Opening his campaign in October 1903, he explained that he saw no reason why he should not stand against a Liberal. He argued that the customary argument about splitting the anti-Tory vote and so letting a Conservative in was invalid in this instance. He pointed out that most of the miners in Accrington were Conservatives as was more than half the membership of the local L.R.C.[8] Accrington Liberals admitted this point, and some of them suggested that Greenall would stand little chance since the miners were strong Tories and would not change their voting habits. Nevertheless, it was felt at national level that pressure should be brought to secure his withdrawal. The Liberal chief whip talked to the Labour leaders. They responded by putting further pressure on Greenall and refusing to assist him in his campaign.[9] At the same time, the more promising constituency of Newton-le-Willows was left open as a bait. In December 1904, the executive of the L.R.C. had to concede that undue pressure had been brought to bear. Within a month, Greenall withdrew but did not contest elsewhere.[10]

The national leadership of the L.R.C. only secured a pyrrhic victory. Local feeling ran high at Accrington over the pressure to which Greenall was subjected. In spite of the disapproval of the L.R.C. national executive, in August 1905 the Accrington Trades Council decided to run Dan Irving, of the S.D.F., against the Liberal candidate.[11]

By the time the 1906 general election came, there had been many changes of candidates and seats. The Lancashire miners were left with two contests—Stephen Walsh at Ince and Thomas Glover at St Helens. Both were fighting sitting Conservative Members for seats which the Liberals did not intend to contest. At Ince, Walsh was fortunate because the election happened to coincide with a strike. Miners in the constituency, their feelings already aroused, did not require much persuasion to vote against Colonel Blundell, the coalowner. At St Helens, Glover's opponent—Sir Henry Seton Karr—was not primarily concerned about politics. According to historians of the town, 'his chief interest lay in a cattle ranch in Wyoming and the hunting of wild animals'.[12] His connection with the constituency came through marriage: he wedded one of the daughters of the Pilkingtons, the glass manufacturing family, who dominated St Helens industrially and politically.

The results in the 1906 general election were:

St Helens		*Ince*	
T. Glover (Lab.)	6,088	S. Walsh (Lab.)	8,046
Sir H. Seton Karr (C)	4,647	H. Blundell (C)	3,410
Majority	1,441	Majority	4,636

Walsh and Glover became the miners' first M.P.s to join the Parliamentary Labour Party. A wave of enthusiasm greeted their victory. Lancashire miners realized parliamentary politics paid dividends. They decided to sponsor three further candidates—H. Twist, T. Greenall and J. E. Sutton in Wigan, Leigh and East Manchester respectively. As Dr Roy Gregory says, the Lancashire miners 'had no inhibitions about contesting seats held by Liberals'.[13] Where their interests conflicted, they expected the Liberals to withdraw or take the consequences. The result at Wigan in the 1906 general election revealed the intrinsic weakness of the Liberals in a Lancashire mining seat. Thorley Smith, the chairman of Wigan L.R.C., pushed the Liberal candidate to the bottom of the poll:

Sir F. S. Powell (C)	3,573
T. Smith (Lab.)	2,205
W. Woods (Lib.)	1,900
Majority	1,368

This result indicated that the miners had little to lose if they fought the Liberals in the industrial parts of Lancashire.

As a result of their own experiences—their own successes and the hostility of the Liberals—the Lancashire miners were politically precocious, prepared to associate with the emergent Labour Party while miners in other coalfields remained reluctant or, at best, lukewarm. This was confirmed in 1908, when a national ballot was taken on whether the M.F.G.B. should affiliate to the Labour Party. Lancashire miners were overwhelmingly in favour of the proposition. The ratio of those in favour to those opposing affiliation was greater than in any other coalfield:

For		*Against*	
Lancashire and Cheshire	2·20	South Derbyshire	0·61
Cumberland	1·85	Midland Federation	0·54
South Wales	1·67	Nottinghamshire	0·50
Somerset	1·59	North Wales	0·41
Yorkshire	1·58	Derbyshire	0·35
Northumberland	1·40	Leicestershire	0·29
Scotland	1·24		

Source: R. Gregory, *The Miners and British Politics*, p. 185.

Labour with a strong tinge of Socialism were the politics of the Lancashire men. Although it is true that all colours of the rainbow found expression among the colliers, certainly the dominant colour was red. Three factors appear to have influenced this development:

First, there was, as shown in the last chapter, the remarkable strength of socialist organizations in Lancashire, which affected the general political climate in the county.

Second, there were the appalling conditions in the mines, worse than most other parts of Britain, with which the L.M.F. failed to cope effectively. In 1912, the *Daily News* painted the grim picture, contrasting Lancashire's discomfort with the comparative ease of other coalfields:

> 'Wigan, the Lancashire coal centre, and Chesterfield, the Derbyshire coal centre, are only fifty miles apart as the crow flies. They might be five hundred if a difference in spirit depended on geography. In Derbyshire the demand for a minimum wage is based on economic theory supported by a sense of comradeship with less favoured districts. In Lancashire, the men will tell you it is a matter of life and death. For years the history of the Lancashire coal industry has been a history of miserable strikes following miserable strikes in weary and often futile repetition, of lock-outs and actions at law. The cumulative effect is an estrangement between masters and men such as exists nowhere else except in South Wales.'[14]

Third, in these circumstances socialist demands had a greater relevance and impact. When the S.D.F. called for a minimum wage, the Eight-Hour Day, nationalization of the mines, it struck a responsive chord in minds of Lancashire miners.

Historians have remarked, quite correctly, that the S.D.F. debarred itself in most industries from fruitful contact with the workers. Its language of doctrinaire Marxism, coupled with a hostility to strike action, effectively built a barrier between it and the people it wished to persuade.[15] Despite these handicaps, the S.D.F. 'greatest achievement', as Dr Hobsbawm pointed out, 'was to provide an introduction to the labour movement and a training-school for a succession of the most gifted working-class militants'.[16]

But what were ideological incumbrances elsewhere in Britain were positive assets in the Lancashire coalfield. The revolutionary language of the S.D.F. echoed many colliers' feelings of frustration and hostility.

Moreover, it should be remembered that, while the S.D.F. opposed strikes, it was not opposed to trade unions. Since many of the struggles in North-East Lancashire where the owners sought to crush the union, S.D.F. members could assist in the dispute with a clear conscience, claiming that they were attempting to establish recognition and negotiating rights. In any case, however, there is strong evidence to suggest that a good deal of local autonomy existed within the S.D.F.: when a branch, particularly if it was the size of Burnley, disagreed with national policy, then it just disregarded it.

The S.D.F.'s official policy of eschewing strikes, believing that industrial action was less efficacious than political action, appeared to make sound sense to Lancashire miners, who were painfully aware of the unfavourable relationship of forces prevailing in most mines. When H. M. Hyndman debated with Charles Bradlaugh, his arguments for the Eight-hour day seemed eminently sensible, and worthy of support.[17] He argued that the trade unions were too weak to achieve this objective by industrial means; therefore, it could only be achieved politically. In the past Parliament had enacted legislation restricting women and children working in coal-mines, and this had generally been beneficial. Similarly, Parliament could go on to restrict hours of labour for adult males, and this also would be for the general good.

Without the socialist phraseology, these arguments had equal appeal to Liberal and Conservative miners. Sam Woods made the same point as Hyndman in 1898 when he said:

> 'Eighty thousand engineers had for six months been fighting by their unions to get the Eight hours day adopted, and their families had been starving and trade had been driven out of the country and yet to all appearances they were no nearer their object than at the beginning. The miners believed they ought to have an Eight hour day, and the best method of getting it was by legislative enactment.'[18]

A similar confluence of people with diverse political opinions occurred over the question of the reorganization of the coal industry. Here there were two distinct but allied issues—mining rents and royalties, the nationalization of the mines.

On the first of these, socialists had continually campaigned. The apparent injustice of the situation was frequently referred to at public meetings in Lancashire. The first of the S.D.F.'s 'Politics for the

People' pamphlets was on Mining Rents and Royalties by J. M. Davidson, who stated his case as follows:

> 'If there is one thing in this world more astonishing than that individuals should claim private property in the surface of the planet, and have their claims allowed by the Legislature of a free country, it assuredly is that they should pretend to have a right to the content of the interior. A coal-hewer descends into the bowels of the cold earth, and with infinite toil and danger raises a ton of fuel for tenpence or even eightpence. Another man, calling himself a landlord, who is meanwhile, perchance, gambling at Monacco or bear-hunting in the Rocky Mountains, successfully exacts a toll of thirteen or fourteen pence a ton on the entire output of a mine, or, it may be, a score of mines! Could there be a more startling anomaly? "O Lord, What fools these mortals be!"'[19]

Of course, miners hoped, if royalties were abolished, it would be possible to advance wages.

But, equally, coalowners might harbour similar ideas for personal gain—an abolition of royalties would leave more money available for profits. The English Land Restoration League, a body expounding the views of Henry George, published a pamphlet by J. E. Woolacott stressed the point that such a change would be in the mutual interest of coal masters and men. The E.L.R.L.'s report for 1886 said this pamphlet was 'in great demand in the mining districts'.[20] In the same year, the T.U.C. passed a motion that wanted 'immediate steps for instituting an inquiry into the character and amount of these royalties and charges, with the object of making them national and not individual property'.

The government appointed a Royal Commission to inquire into Mining Royalties in 1890. Spokesmen for the Lancashire coalowners, Messrs Hewlett, Blundell and Grimke, spoke of the grave hardship they occasioned, particularly in times of depression. Nevertheless, they did not favour nationalization. Nor did Thomas Aspinwall, who was one of the union's representatives.[21]

But Sam Woods, the other representative of the Lancashire Federation, took a different line. He thought that, at the most, royalties should be twopence a ton:

> 'I say that if royalties were reduced well down to a minimum price, I believe there would be a sufficient margin to allow greater profits than the employers are at present receiving, and there would be greater wages.'[22]

Woods went on to advocate more sweeping changes:

'I hold this view: that this nation is very much handicapped by the present system of leasing mines; that is, the lease being in the hands of the private owners; and I think that the coal industry, and probably the iron industry as well, suffers seriously on this account. I believe if the State could get these royalties into their possession—get the mines into their possession—they would be able to deal with private leases far more satisfactorily than they are dealt with at present.'[23]

While Woods favoured nationalization of royalties, it is important to realize that he did not see this as a step towards the socialist millennium but as a means of making existing class relationships more harmonious. He mentioned that Lancashire owners often gave the high level of royalties as a reason for not granting wage increases. When trade was bad, royalties could make that marginal difference that led to pit closures. Also, it was a source of industrial strife; without them there could be greater mutual accommodation by masters and men of each others claims. He cited the strike at J. & R. Stones, of Garswood, where a hundred men had been out for six months for a penny or twopence a ton.[24]

The Scottish miners' spokesmen before the Royal Commission on Mining Royalties went further than Sam Woods. Robert Smillie and William Small came out unequivocably for nationalization of the mines.[25] From then on, the Scottish miners were to the forefront on this issue.

But their Lancashire brothers were not far behind. It was they who brought the resolution before the 1894 M.F.G.B. conference which suggested that 'the best interests of the nation will be served by the nationalization of the mines of the country'. Tom Greenall moved the motion. What followed is not recorded either in the annual conference report of the M.F.G.B. (which up to this point had been reporting conference proceedings fully) or in the newspaper accounts. But they all agree that, after 'a long and animated' debate, the pro-nationalization resolution was carried by 158 votes to 50.[26]

Once the idea had been accepted by the M.F.G.B., then the question of its wider application was raised. One of intricate problems of the Lancashire coalfield was the maze of mining concessions, resulting in difficulties arising because, to get the coal, one had to cross another person's land. So it seemed logical to add land nationalization to that of coal. But then there was another problem that plagued the industry in Lancashire—namely, the failure of the railways to provide the required service—and hence there was a further candidate for nationalization. Consequently, the Hindley district brought this plan for more comprehensive public ownership—land, railways and the mines—before the M.F.G.B. annual conference the following year, 1895.

The executive of the M.F.G.B. ruled the Hindley resolution out of order. It was argued that it was not specifically related to the coal industry, brought in matters of an extraneous character, and violated the union's traditional position of being non-political. Significantly, the Lancashire Federation opposed the national executive's ruling, a sign that there must have been widespread feelings on the issue throughout the county.[27]

But, perhaps, more interesting still is the Hindley district union itself. It seems to have been, at least until 1892, a Conservative stronghold. The election of Sam Woods that year as M.P. for Ince led to widespread resignations. The *Wigan Comet* reported, 'In Hindley District alone over four hundred severed themselves from the old organisation', and many appear to have joined the Conservative Labour Union.[28] For a time, membership of the Hindley District was reduced to a mere three hundred. But by 1895, membership had again risen to the six hundred to seven hundred odd mark. While this could possibly be the result of an influx of new miners, a more probable explanation is that, after the demise of 'the blue button union', many men returned to their former allegiance. If this latter theory is correct, it means that union that called upon miners throughout Britain to favour extensive nationalization had a considerable proportion, perhaps a majority, of Conservative members.

Be that as it may, the Lancashire Federation refused to remain content with the M.F.G.B. executive's ruling, and it was eventually reversed. In 1897, at the national conference, Lancashire moved that it was 'essential for the maintenance of British industries to nationalise the land, mines, mineral royalties and railways of this country'. It was carried by ninety-seven thousand votes to six thousand. But, to some extent, its effect was lessened because conference also passed a resolution from Yorkshire, stronghold of Ben Pickard and the old brigade, which called on 'all congresses to act on trade union lines as in the past and not on socialist lines'.[29]

Of course, it was not only some of their fellow miners who jibbed at the politics implicit in the demands of the Scottish and Lancashire men. To the two political parties, the package of demands was unacceptable. This or that point might be partly conceded, but in their totality they were not reconciliable with traditional Liberal and Tory principles. So the Lancashire union was confronted with the agonizing position of having no alternative, if it wanted to achieve its objectives, but to move towards Labour politics.

A paradox of the Lancashire miners was that, although they possessed wide disagreements about political form, there was almost unanimity about the political content. Disputes were over what party to support; not what should be its policy for the coal industry. Miners might be fervent Tories, Liberals or Socialists, but they all seemed to want the legislature to do the same things—introduce stricter inspection, enact more stringent safety legislation, give them the Eight-hour day etc. Indeed, as we have seen, the achievement of these demands was regarded as almost a matter of life-and-death.

So, consequently, Lancashire's Liberal and Conservative miners found themselves with a profound inner conflict—their basic economic needs conflicted with their political loyalties. Ultimately, economic necessity proved to be supreme. An examination of the two political parties show why this happened.

An analysis of Toryism among the Lancashire miners show that it contains political attitudes that can be grouped together under three different headings. First, there were those who voted Conservative out of deference. The reasons that produce the deference voter in a hierarchical society have been explained by political commentators.[30] They change gradually from age to age, but such small changes cannot satisfactorily account for a large scale shift in voting habits. Second, there were historical reasons—a hatred of the grasping Liberal mill-owners during the early phase of the Industrial Revolution, a gratitude to Conservative governments which enacted the factory acts and other progressive social legislation. Insofar as these historical influences were not confirmed by contemporary experience of a similiar kind, they tended to represent a dwindling capital asset of the Tory Party, a fund of goodwill that would be exhausted as memories faded. Third, there were a cluster of political attitudes—fervent belief in the monarchy, the British Empire, one's country and its military might—that were extensively held among miners and found their most forceful political expression through the Conservative Party.

But chauvinism, in this period, was far from being the exclusive property of the Right. The Left, too, had its flag-wavers. Robert Blatchford called for military training for all boys over the age of ten.[31] He thought that increased military expenditure an absolute necessity: 'We must have an Army and an invincible Navy.'[32] It is alleged that his daughter played 'Rule Britannia' to him on the piano

(a) Leigh businessmen and traders had a medal struck to commemorate The Battle of Howe Bridge. One was presented in gratitude to the commanding officer of the cavalry.

(b) A pair of Lancashire miner's night clogs.

Coal gathering during the miners' lock-out of 1893.

every evening all through the Boer War.[33] H. M. Hyndman likewise was a believer in military might. In 1910, during the hysteria over German dreadnoughts, he joined the Tories in asking for a further one hundred million pounds to be spent on the navy.[34]

Since this was an age in which socialism could be combined with chauvinism, it was made much easier for people like the Lancashire miners to make the transition from being supporters of the Tory Party to joining, say, the S.D.F. Indeed, in so doing, they would merely be repeating the political evolution of Hyndman himself, who moved from being a Conservative to being a disciple of Karl Marx.

The Lancashire miner's personal position, the pressures of the environment and the need for legislation to improve their lot, led to Conservative attitudes becoming anchored to a new political allegiance. The Tory Party had deep roots in Lancashire—there were many Conservative working men's clubs—and when the change came it often involved swinging quite far to the Left.

For Liberal miners, the adjustment was less dramatic. Throughout the country, in every other coalfield, colliers were making the transition from Liberalism to Labourism. Probably of less numerical strength than the Conservative colliers, they appear to have usually been more moderate politically.

Even so, many Liberal miners, like their Conservative counterparts, were receptive to the new ideas, ready to re-appraise their voting habits. It was a time of flux, of change, when many minds were open to be convinced. Perhaps this is best shown by what happened when, during the formation of the I.L.P., there was a move to exclude from its ranks people who belonged to Liberal and Conservative clubs. The secretary for Lancashire wrote to the *Workmen's Times* opposing the suggestion since it 'would cut away our best recruiting ground'.[35]

CHAPTER FIFTEEN

The Lancashire Miner and his Social Life

The manner whereby individuals earn their livelihood exercises a pervasive influence on all aspects of their lives. Miners, inured to rough work with no provision for comfort or sanitation, naturally demand none too exacting standards when they come to the surface. They tolerate living conditions which would be anathema to other types of worker. In 1860, Dr E. H. Greenhow vividly illustrated this point. He says that when colliers were sentenced to jail with hard labour, they accepted their fate stoically; the only complaint they made was that they were obliged, whilst at work, to keep the ordinary posture of rational creatures'.[1]

Lancashire miners in the first half of the nineteenth century complained very little about their living conditions. Even if they had, it seems unlikely that the coalowners would have improved them. They had no pressing reason to do so. An absence of local pride resulted in scant prestige being derived from the introduction of improvements. Many coalowners, anyway, had no close associations with the district where their mines were situated. They usually lived in pleasant country-houses, well away from their collieries. Their attitude was akin to that of absentee landlords, wanting to extract the maximum of wealth at the minimum cost. They saw nothing was to be gained from squandering money, pampering the collier and his family. A 'single cottage'— that is to say, a cottage with only one room—would suffice to meet the colliers' modest needs.

The nature of life in Lancashire made it difficult to make lasting improvements. In the early stage of the Industrial Revolution, the county became enveloped in a smoky squalor that people came to regard as part of the natural order of things. The reason for this lay in the comparatively large consumption of coal. Besides the expanding demand of the growing population for domestic purposes, industrial needs also grew inordinately as more and more steam-engines were employed. Of all industries, textiles installed the greatest number. But they were also extensively used on canals, in mines, and by breweries, all of which were to be found within the county precincts.[2] By 1838, towns like Manchester and Salford possessed almost three times as much steam-power as Birmingham.[3] As a result of this intensive industrial development, Lancashire people became habituated to grime and dirt.

The unplanned and rapid expansion of the North-West, with the economic interests of capital overriding considerations of human welfare, served to exacerbate the plight of the people. In 1841, 102,025 Lancastrians died, 83,616 of them under the age of 20. The average age of death was 22·1 years.[4] Lyon Playfair, who gave these figures in his report, said crowded cities produced 'large quantities of animal and vegetable refuse'. Many houses, grouped around court-yards had open cesspools, dunghills or middens. He calculated that there were sewage heaps of between twenty-five thousand and thirty-five thousand tons in many Lancashire towns.

A horrifying picture of squalor, dirt and disease emerges from official reports published in the early 1840s.[5] Independent investigations tend to confirm the view that living conditions were extremely bad. According to Frederick Engels:

> 'Among the worst of these towns after Preston and Oldham is Bolton, eleven miles north-west of Manchester. It has, so far as I have been able to observe, in my repeated visits, but one main street, a very dirty one, Deansgate, which serves as a market, and is even in the finest weather a dark, unattractive hole in spite of the fact that, except for the factories, its sides are formed by low one and two-storied houses. Here, as everywhere, the older part of the town is especially ruinous and miserable. A dark-coloured body of water, which leaves the beholder in doubt whether it is a brook or a long string of stagnant puddles, flows through the town and contributes its share to the total pollution of the air, by no means pure without it.'[6]

Matthew Arnold seems to have had a slightly different opinion from Engels about the worst Lancashire town: 'St. Helens is eminently what Cobbett meant by a *Hell-hole*.'[7] However, for Ruskin, out to find evidence of the ugliness produced by industrialism, Wigan epitomized the evil of the system. He found the vision so horrific that he never ventured further than the station. In his book, *Fors Clavigera*, specially written for workingmen, he described the scene:

> 'Those children rolling on the heaps of black and slimy ground, mixed with brickbats, broken plates and bottles ... the children themselves, black and in rags evermore, and the only water near

them either boiling or gathered in unctuous pools, covered with rancid clots of scum, in the lowest holes of the earth-heaps ... Are they not what your machine gods have produced for you?'[8]

If one accepts the death rates as the nearest approximation to being an objective criterion, Wigan was the most unhealthy town. The Board of Health regularly issued statistics for 50 towns in England and often Wigan secured top place. In 1875, Wigan's death rate was 42·1 per one thousand whereas the national average was only 27·5. Since Wigan, at the same time, also had the highest birth rate—53·2 per thousand as against a national average of 33·5—the town's high mortality figures cannot be attributed to an aged population.[9] In 1873, Netten Radcliffe, of the Local Government Board Medical department, itemized the town's main defects: want of sewers and drains, want of paving and channelling, bad construction, ventilation and dirtiness of houses, want of water, lack of refuse disposal.[10]

Although it has been customary to concentrate attention on Lancashire towns like Wigan, it seems probable that even worse conditions prevailed in many of the surrounding mining villages. Yet, these tended to be overlooked and were rarely visited by investigators, who concentrated on the main centre of population. Since mortality figures were usually not given for each village—they were included in the overall figures of larger districts in the county—it may never be possible to prove this point.

Even so, some evidence does exist. In 1874, *The Lancet* reported the Local Government Board inquiry into Skelmersdale, 'a colliery village having a population of four thousand':

> 'The statements made in evidence as to the condition of the village show that, even among colliery villages, Skelmersdale must have a pre-eminence in filth. The houses, many of the most miserable construction, are almost buried in the filth of the inhabitants; the privies so foul as to repel even those most familiarised with them; the drainage accumulated in horrible puddles, fed also by the liquid abominations of pig sties and midden-steads.'[11]

A diptheria epidemic was then in progress in Skelmersdale. In nearby Upholland, Dr L. Molyneux, the medical officer of health, reported about houses with open gutters for sewerage to flow along and cesspools seeping into pantries: 'I was asked if I considered it was not enough to make a dog sick, and it was certainly vile.'[12] At another village—Aspull—William Clark, the nuisance inspector, reported that in 'several cases where cattle have died from disease, I have prevented the carcasses being used for human food'.[13]

The Village of Crooke's Complaint, published in the *Wigan Observer*, gives some idea of the conditions still prevailing in 1874:

> 'I do not think there is a more dirty and more neglected place in England than I am ...
>
> 'I am told that in other places the people who live there put a top-coat of stones on their streets, and so when they want to walk about they can do it and have their feet dry even after a heavy shower. But with me its different; people build houses on me and leave so little space between row and row that the air can scarcely get between the rows, and the roads are made of dirt, and when people walk about on them they take a good deal of dirt into their house when they go in, and no matter how clean the wife is the house is just as if it never had been cleaned, with the husband and children going out and in with dirty clogs, and so the insides of the houses, I am told, are always dirty. And I don't think the houses are ever whitewashed, or not very often, so when sickness comes it lays a terrible hold on the people, and they can scarcely shake it off.
>
> 'I heard somebody read a letter out of your paper about the dirty state I am in, and the letter said there was scarlet fever and other zymotic diseases about the place. I cannot tell what zymotic diseases are, but if they have anything to do with dirt I can well believe the writer tells the truth. I have not got a decent drain through me, and whatever dirty water or slops are thrown down they sink into me all over and make me quite ill.
>
> 'I hear that this authority is going to burn a deal of property and the fever with it, and they are going to close the schools, and the children won't have any place to go to, and then they will go into each others houses and spread zymotic diseases, and then there will be more burning property. It is a pity to make such a loss to poor people; don't you think it would be better to prevent than cure, as I think they might have done?
>
> Crooke Village'[14]

Miners' social life was inevitably affected by living amid squalor, dirt and degradation. While mental improvement societies and other educational bodies endeavoured to enlighten them, such activities appear to have been very much a minority interest. Most men seem to have neither the time nor inclination. Their main pastime appears to have been drinking, often heavy drinking. Coalowners frequently complained that the heavy consumption of alcohol resulted in a high absentee rate; even those who stumbled to work were, because of their condition, a danger to their own safety as well as that of their fellow workers. A union leader criticised men 'whose minds never soared any higher than the brim of a beerpot'.

The pleasure of drinking was augmented by such traditional miners' games as whippet-racing and cock-fighting. The latter appears to have been popular in the first half of the nineteenth century: there was a saying that the Lancashire collier's coat of arms should have been 'a stark naked child and a game cock'—a symbol of the large sums of money lost gambling on cock fights.[15]

There were, however, rather more unusual pursuits. The *Wigan Observer* reported a linnet-singing match at a public house at Goose Green where, it announced, 'a copper kettle and other prizes will be given'.[16] The entrance fee of a shilling entitled a competitor to sup sixpence-worth of ale. Judges decided which was the best linnet by the quality and quantity of sound produced; some birds chirped continuously for five seconds. Then there was a variant of a beauty contest, known as 'a pastry feast'. Usually held at public houses, young courting couples were invited to parade in front of the assembled drinkers, running the gauntlet of ribald comment. The customary prize for the most attractive girl was a pint of rum and a sixpenny pasty while the ugliest female competitor received a pound of black pudding.[17]

A much less harmless pastime was 'purring'. This was a game of football played without the unnecessary complication of using a ball. Two men just kicked each others shins until one stumbled and was then battered into insensibility. A St Helens pitman explained 'the game':

> 'It is all up and down fighting here. They fought quite naked, excepting their clogs. When one has the other down on the ground, he first endeavours to choke him by squeezing his throat, then he kicks him on the head with his clogs. Sometimes they are very severely injured.'[18]

'Purring', like boxing, was done for money. Matches were watched by spectators, who bet heavily on the result. But besides being regarded by some people as a sport, it also had two other functions: first, as a means of inflicting grievious bodily harm—or even killing—an adversary in a quarrel; and, second, as a means of showing displeasure with one's wife and children. As the *Manchester Guardian* remarked, 'a good deal even of the strictly local kicking takes place within the charmed circle of the home and is never allowed to be the topic of public scandal'.[19]

Notwithstanding this fact, sufficient examples of 'purring' did come to the authorities' notice to occasion them grave concern. In 1873, the coroner of Wigan wrote to the home secretary, informing him that four deaths had occurred in the borough due to 'purring' during the past two years: 'The clogs worn by the working classes here have wooden soles with narrow toes bound with iron and tin; they form most dangerous weapons, and are readily resorted to on the smallest provocation.'[20] Commenting on this letter, the local newspaper explained that, on average, a man was killed or maimed every fortnight through the practice.

Yet, it was difficult to stamp out this dangerous and barbaric pastime. Supporters of 'purring' argued that it always had been done and, therefore, always should be done—'the clog is the traditional Lancashire method of duelling'. They also claimed it was less dangerous than other forms of duelling, like 'my Lord Tom Noddy's pistol or sword'. Furthermore, the law would be showing an upper-class bias if it dealt severely with men found guilty of 'purring' and, as was proposed, flogged them while 'gentlemanly barbarians are not dealt with in that way'.[21]

Some idea of the magnitude of the problem can be seen from this incomplete list of injuries in Lancashire due to kicking during a typical month. The list is incomplete because (1) all newspapers were not inspected, (2) 'purring' matches are not included, and (3) many instances would not come to the notice of the police:

August, 1874

August 1:

Oldham: Two men kick a third to death; five years' penal servitude each.

Oldham: Workmen, passing dinner-hour at an inn, struck an old man. A passer-by shouted, 'Shame' and consequently was kicked to death. Two men sentenced to eight and ten years' imprisonment.

St Helens: Six colliers on the spree at Parr attacked several cottages, smashing windows and doors. They entered the house of an eighty-year-old man, kicked his eyes out, forced lime down his throat and also filled his bleeding eye sockets with lime.

August 3:

Oldham: Two men killed a stranger after a pub brawl. Victim's head fractured. One man sentenced to fifteen years' imprisonment, the other acquitted.

Liverpool: Roughs asked man going home with his family for sixpence. When he refused, they kicked him to death. They had a long knife but preferred to use their feet.

August 6:

Bolton: Brawl in Recreation Tavern. One man kicked another's shins. Whereupon the second went home, borrowed some clogs, and retaliated by kicking the first man severely in the stomach. Two months' in prison.

Bolton: One month in jail for kicking a constable. The policeman said he was obliged to hit the man on the head with a stick since the man was 'kicking so fast'.

August 8:

Duckinfield: A collier kicked a woman to death. Received five years' penal servitude.

Stockport: A man kicked to death a woman who lived with him. Seven years' penal servitude.

August 11:

Bury: Three men kicked a man to death in a public house.

Aspull: A man kicked to death by his workmates.

August 14:

Westhoughton: Sixty-year-old woman kicked to death by her husband, who had been drinking.

August 15:

Wigan: Passing a ten-year sentence on a young Wigan collier, who attacked an eighty-four-year-old man, Mr Justice Archibold said, 'It seems to have occurred to Glover that he would kick Derbyshire, and so he did; but entirely without malice, for when he was quite done with the old man, he commended him to the care of the matron in the house, telling her that he had been "purring Daddy" and was afraid he had done him some harm. Daddy was found lying in a pool of blood, which had run from two wounds in his head averaging three inches and a half in length and penetrating the bone.'

August 18:
Aspull: Three men attacked a fourth, kicking out his teeth.

August 21:
Manchester: A man kicked a woman with whom he was living. He had been convicted on three previous occasions for the same offence. Imprisoned for six months.

August 23:
Ashton-under-Lyne: A man 'purred' for the 'price of a quart'. He was also turned upside down, and two shillings and eightpence stolen.

August 26:
Preston: A six-year-old child kicked and paralysed in the leg.

In many of the towns and villages where people seem to have been most prone to resort to physical violence, the police were weak, both numerically and in the calibre of their personnel. In his study of the Lancashire police force, Mr E. C. Midwinter states that, of the county constabulary's first two hundred recruits, fifty were discharged within the first six months. The casualty rate was even higher at Wigan, where the Watch Committee minutes show that policemen assaulted and arrested each other, made false expense claims, stole from a murderer, made water on the police station floor, slept during a riot, and found the Wigan Brewing Company premises too tempting to leave alone. The entire membership of the Wigan police force, except for the chief constable, had been dismissed in a little over a year.[22] Yet, the chief constable's turn was to come: he was sacked for drunkenness and misconduct while on duty. 'I thought it was not proper of him,' said Martha Seddon, 'To put his hand on my bosom drunk.'[23]

In the riots of November 1853, the Wigan police played an ignominious role. When the crowd turned angry, the police retreated into its barracks, locked the door and left the Mayor and other civic dignitaries on their own to face John Leadbetter (known as 'the Poker') and his confederates. (See Chapter Three for fuller account.) The riots cost the town £2,066, including £1,141 for the maintenance of soldiers required to quell the disturbance. Wigan public hall, police station and grammar school had all been damaged.[24] Besides the public expenditure incurred in restoring these buildings, the citizens of Wigan themselves had to pay for repairing their own private property. This must have fallen heavily upon the wealthier section of the community. The value of property in the town, as Table 19 indicates, was becoming considerable, and it was imperative to protect it. After 1853, Wigan Council saw to it that the town had a police force of sufficient size and calibre to preserve law and order. The cost of doing this was far less than the threat to private property consequent on not doing so.

Enlightened self-interest of the richer people in Lancashire, those with the power and political influence, appears to have been the main motivating force that led to improvement generally. The same argument that applied to the Wigan police could be used with equal validity elsewhere. Unhealthy urban conditions constituted a threat to the industrialist's life and to the profits of his business. As long ago as 1837, statistics showed that there was a common bond of suffering between the master and his servant—both had their life-span reduced, albeit by different amounts, as a result of living in towns without adequate public health provisions:

	Average Age of Death	
	In Manchester Years	In Rutland Years
Professional persons and gentry, and their families	38	52
Tradesmen and their families (in Rutland, including farmers and graziers)	20	41
Mechanics, labourers and their families	17	38

(Source: Manchester Statistical Society.)[25]

In 1841, Lyon Playfair estimated that fourteen thousand of the 102,025 deaths occurring in Lancashire had been preventable. Similarly, there was in that year three hundred and ninety-eight thousand cases of sickness, costing £5,133,557, that could have been avoided.[26]

As industrial production increasingly required a smooth flow of inputs, the effect of epidemic—sudden absences of large numbers of the labour force—must have become more and more irksome to the employer. Likewise the death of a skilled worker in his prime meant that somebody else would have to be trained to take his place. This usually involved time and trouble, during which total output would consequently be less.

Enlightened self-interest of the employing class—or, at least, a large section of it—was the principal force responsible for ameliorations. Yet, it would be wrong to attribute environmental changes to simply one influence. Throughout the county there were many forces at work which served to make people call for higher standards, regarding what was tolerated yesterday as intolerable today.

Within mining communities, a number of factors led to a transformation of outlook. Probably the most important of these arose from the miners' own practical experience. In part, this resulted from the growth of communications and hence a more frequent contact with other people who had different, and usually higher, standards. Occasional visits to seaside resorts and cities in other parts of the country must have given miners some inkling that their living conditions could be improved. But more important than the impact of such fleeting experiences was that of their changing social situation. The growth of towns in Lancashire promoted occupational de-segregation, thereby providing conditions in which the uncouth could acquire more civilized ways. Samuel Bishop, a St Helens glass manufacturer, described the effect of this process on miners:

> 'Colliers have a bad name; they used to live as it were isolated from everyone, and were rough and ignorant: to some extent they are so still but towns have grown up around coalpits of late years ... The colliers have now been forced to mix more with their fellows and have improved accordingly.'[27]

Then there was the effect of universal education, which helped to widen mental horizons and inculcate new social values. This was, perhaps, slow in coming to some parts of Lancashire as many colliers thought no tangible benefit was gained from schooling. Their doubts were echoed by Peter Higson, the mines inspector. He wrote in his 1870 report: 'To keep young persons from work till they are 12 years of age will, I fear, create the objection to labour which through their life they will be unable to overcome.'[28] With such opinions prevalent, it is not surprising that, in 1875, the Rev. T. F. Fergie, of Ince, discovered that most colliery boys received no education: of an estimated 2,500 who should have attended in his district only 1,044 in fact did.[29] Truancy also seems to have been common in other Lancashire mining areas. Nevertheless, compulsory school attendance was eventually enforced. It, undoubtedly, had a profound beneficial influence in the long run.

The various miners' unions also played their part.[30] The more percipient leaders realized that, if the men were to attain a better life, then they would have to develop new attitudes. Drunkenness and profligacy were deplored; fighting was regarded as unbecoming. They had to replace these bad habits with self-respect and respect for their fellow men. Insofar as the unions helped to engender this feeling, creating a spirit of friendly collaboration among workers, then they were proud of their role. For example, in 1845 a union dinner at Little Lever led the *Miners' Advocate* to comment:

> 'Al passed off as meery as "marriage bells", notwithstanding the prognostications of some that miners could not meet at a convivial feast without a fight. But they knew better now; the union has knitted them together, made them better neighbours, better husbands.'[31]

The greatest advance, however, did not come from miners being husbands so much as pit-lassies becoming better wives. When women worked down the pits, doing physically exhausting labour for long hours, they did not have the time and energy to keep their homes in a decent condition. The introduction of the 1842 Act of Lord Ashley led to a gradual improvement

It is, perhaps, sometimes not appreciated to what extent the problem of female labour underground was a localized problem, confined to Lancashire and a few other areas. The 1841 census stated that there were 118,233 coal miners in Britain, of whom only 2,350 were women. Roughly a third of these women—between seven hundred and eight hundred, according to the mines inspector's report—were employed in Central Lancashire.[32] Testimony to the 1842 Royal Commission confirmed the fact that the custom tended to be confined to a comparatively small district: Wigan colliers who went to work in Manchester mines returned to their native town, complaining that no women worked underground and they were expected to wash more frequently.[33]

It is only possible to speculate about the reason for this use of female labour. Perhaps it has something to do with textiles—under the domestic system families worked together as a unit. Likewise, it has been shown that the Lancashire agricultural labourer was usually helped by his wife and family.[34] It may be that this tradition of family toil was ultimately transferred to the mines: the husband hewed the coal while his wife dragged it out.

Whatever the explanation for women working underground, when the practice was discontinued it aroused considerable opposition. Two hundred women were illegally employed in Wigan mines three years after the 1842 Act had been passed. Coalowners placed lookouts to keep on the watch for approaching policemen and mines inspectors. They also dressed women up in men's clothes in an attempt to avoid detection. 'One of the collieries at which the Act has been notoriously evaded, by having ladders openly placed at the pit-mouth, belonged, I regret to say, to one of the borough magistrates,' declared inspector Dickinson in his 1845 report.[35] Prosecutions ensued but, as his 1846 report shows, it did not stamp out the practice:

> 'I regret to say that, notwithstanding these convictions, there are coalowners in the district around Wigan, and in that part of Lancashire, who continue to violate, or connive at the violation of the law ... Police cannot give up the whole of their time to this matter ...'[36]

Nevertheless, the Act ultimately was enforced. Women who had worked underground found other jobs, becoming full-time housewives, working in the mills, going into service, or even continuing at the colliery sorting coal on the pitbank. Among the more bizarre examples of industrial redeployment took place at Winstanley, where women were trained as engine-drivers, driving coal down to Wigan pier. The company's accounts show they were none too successful: they knocked down and killed a small child (compensation £1 10s.) and then, a few months later, struck a cow crossing the line (compensation £10).[37]

Domestic service was a less hazardous occupation. The 1846 mines inspector's report stated that special schools had been opened to equip girls who had hitherto worked in mines for a new employment. Such an establishment was the Walkden Moor Servants' School, a boarding school which took girls between the age of 14 and 20.[38] The students were kept under constant surveillance, right from 5.30 a.m., when the day started with a short prayer, till final prayers at 8.15 p.m. They were not permitted to leave the premises, except if accompanied by a minister of religion, and no males, other than the father, were allowed to visit the girls.

Each new student received an elaborate set of rules that emphasized such virtues as hard work, thrift, honesty and, above all, obedience. Every rule was buttressed by a Biblical quotation, the most ominous of which came from Proverbs xix: 'Slothfulness casteth into a deep sleep; and an idle soul shall suffer hunger.' Any infringements could delay a girl's departure from the school. All the students had been given the same warning: 'If any of you break these rules, the matron will note your name and the offence in her book; and, according to the character in her book, you will be first or last to be recommended to situations in service when they offer.'

Walkden Moor Servants' School was definitely run along Spartan lines. But, nevertheless, it did perform at least one useful function: it introduced new, better domestic customs into mining communities. Higher standards came in from outside.

There is no doubt that, during the Nineteenth Century, there was an improvement in the living conditions of Lancashire miners. Towns were gradually made more sanitary; the provision of schools became more adequate; the children were better fed. Yet, despite these, and many other changes, there was remarkably little change in the way miners were viewed by the rest of the community. They were still regarded with apprehension, a misgiving that arose from the public image of the collier.

The 1842 Report described the Lancashire colliers as 'an inferior class' of persons.[39] That some people still continued to hold this opinion thirty-three years later is obvious from this comment in the *Manchester Courier* of 1875:

> 'We support little crowds of missionaries abroad and yet we have in our midst an amount of vice, profligacy and brutality which would astonish even a South American Indian. In all our large towns modest women and respectable men are shocked at the exhibition of open shamelessness in the streets; thieves and desperadoes prowl about in every highway and byeway; and it is necessary for the decent to be perpetually on their guard. In some villages, especially colliery villages, there is a depth of ignorance and lowness of moral tone which are perfectly astounding ...
>
> 'Last week two men and a woman went to the office of a registrar at Wigan. One of the men was husband to the woman and the other had agreed to buy her for four shillings and sixpence. The money was paid before the astonished official, who was requested to let them "Sign their hands to a bit of paper", a proceeding which in their ignorance they believed would legalise the transaction.'[40]

The same revulsion was manifest in 1890. Baron Huddleston, a judge at Liverpool assizes, used the tragic death of an Ince girl to express his feelings. He said 'that some of the wild inhabitants of that part of the country, the colliers, almost approaching the condition of beasts, had gratified their brutal and lustful passions on the girl, and then thrown her down the pit'.[41] Naturally, Lancashire miners were incensed at being likened to beasts simply because of the misconduct of a handful of their number, and so their union protested to the judge.

Notwithstanding this, the Lancashire Miners' Federation were unable to allay the fears of some sections of the respectability. When, in 1890, it decided to hold the first miners' gala, the people of Southport remained apprehensive. Thomas Ashton reported, 'There were a number of the residents of Southport who were afraid and left the town for the week-end. They said that so great a crowd of rough miners would certainly do great damage in the town, and they were afraid that their very nice recreation ground would be destroyed and ruined.'[42] Their fears, of course, proved to be without foundation.

Prejudice against the collier in the Nineteenth Century is, in some respects, similar to racial prejudice today. The general public do not understand behaviour patterns that are markedly different from their own. They show their displeasure at the heavy drinking, rowdiness and violence of miners; they do not see that these result from the arduous and hazardous nature of their occupation. They do not, moreover, see the very positive qualities that miners possess—their comradeship, their willingness to assist the less fortunate members of their communities, and the courage displayed in times of adversity, such as strikes and pit-disasters.

Nevertheless, it must be admitted that, while there have been improvements, those parts of Lancashire associated with coal mining, still tend to be grim. They still are more dirty, squalid and impoverished than elsewhere. Less social capital appears to have been invested in roads, schools and amenities. St Helens and Wigan spend less on education than any other county borough in England and Wales.[43] A Labour Party pamphlet, published in 1969, lists thirty towns where social deprivation and inadequacy exists: twelve are in the North-West.[44]

The ravages of neglect, dating back for almost two centuries, remains to plague the living.

Table 19 *Principal Landowners around Wigan in 1876*

		Acres	*Est. Gross Rental*
1	Ince Coal and Cannel Co.	6	£7,242
2	Nath. Eckersley	453	£4,967
3	Ryland & Cross	72	£4,731
4	Exec. of T. Morris	40	£4,036
5	Henry Woods	49	£3,397
6	H. Woodcock	88	£3,250
7	Exec. of T. Part	152	£2,648
8	Rev. Bridgman	50	£2,169
9	John Taylor, of Newton	167	£1,729
10	John Speakman	49	£1,617
11	Wigan Corporation	1	£1,450
12	Pearson & Knowles	—	£1,423
13	Moss Hall Coal Co.	—	£1,354
14	M. Pearson	12	£1,023

Source: *Wigan Observer*, 18 February, 1876.

MINERS'

LOCK-OUT.

TUNE—MARBLE ARCH.

YOU Miners all attend awhile,
To please you I will try,
On this great Demonstration day,
Of strength and unity.
You've come here to demonstrate,
Your grievances and the wrongs
The Masters daily heap upon you,
So listen to my song.

Chorus—

Then stick together all
Let no dissention sever,
Your friendship, your rights,
Your Union, good and strong,
The Masters cannot do without
Their Collier lads so clever,
So stick fast to the Union,
And you'll find you won't be wrong.

Thousands of Miners they have come,
With flags and banners gay,
To the good old town,
And their bands of music play.
Just let the Masters look upon,
This motley throng I say,
And see the strength of union,
With the Miners here to-day

Good reason the Miner had of late,
To grumble and complain,
And put up with reductions,
While the Masters had the gain.
But not content, the Masters they,
Seem all quite fully bent,
To reduce the Miners' wages, yet
Twenty-five more per cent.

The Miner dearly earns his bread,
Great dangers he must brave,
Many fathoms down below the ground
It often proves his grave.
Yet with willing heart and hand he goes
To work for loved ones dear,
Then never lower his wages,
Or their union you might fear.

May every demonstration,
Be as brilliant as this,
In brotherly love and friendship,
In harmonious social bliss.
And may our leaders prosper well,
Who so nobly spoke to-day,
On the Miner and his interest,
Welfare and Unity.

Strowger & Son, Printers, Clarence Yard, Wallgate, Wigan.

A ballad produced during the miners' lock-out of 1893. It clearly explains the men's case.

Women working on the pit-brow at the Meadow pit, Wigan, around the turn of the century.
(By courtesy of *Wigan Observer*)

CHAPTER SIXTEEN

At the Turn of the Century

At the end of the nineteenth century, the Lancashire coal industry had developed ominous symptoms, signs of sickness that blighted its further growth. This was caused, first and foremost, by insurmountable physical problems. As surface seams became exhausted, it became necessary to dig deeper. Some of the deepest workings in Britain—at Pendleton, for example, 3,483 feet—were in the Lancashire coalfield. Obviously, this necessitated extremely expensive excavations. To endeavour to offset the cost, owners worked as many as ten seams from the same shaft. Even so, this saving was more than counterbalanced by numerous geological faults, which made operations increasingly difficult and costly.[1]

Deep mining involved a certain logic of production if it were to be accomplished with efficiency. Coalowners needed very much more capital; all resources had to be utilized in a regular, systematic manner; a greater working area was required. At this point, coalowners encountered a second obstacle. The highly variegated ownership of land frequently compelled coalowners to develop in ways which were not the most advantageous. A company like Wigan Coal & Iron could have to acquire as many as fifteen leases to mine one square mile. When somebody declined to have his land used for coal mining, it would have exceedingly harmful repercussions, preventing the systematic extraction of coal and even, in some instances, endangering all workings in the district.

John Knowles, of Pendlebury, told the Royal Commission on Mining Royalties, 1890:

> 'I know of an instance where a landowner in East Lancashire would not grant wayleave or lease the coal. The consequence was that other coal had to be worked near the outcrop, and caused a great inundation of water. This not only drowned out the particular colliery, but stopped the working of adjoining collieries. The remainder of the mine, over a large area, was abandoned, and probably will be lost forever.'[2]

Knowles went on to emphasize the difficulty occasioned 'to work and ventilate the mines when some of the estates have to be avoided'.

But, besides creating problems underground, there could be equally onerous ones on the surface. Coalowners frequently had difficulty in transporting their product. This point was forceably made by the employers' representatives, the Lancashire & Cheshire Coal Association, also in evidence presented to the Royal Commission on Mining Royalties:

> 'In many cases the owner prohibits any but his own coal being carried over his land; in other cases, where access to a railway or shipping place can only be had through intervening lands, the owner of such intervening land exact exorbitant terms, which must be submitted to by the mine worker, or the working of the colliery is seriously injured and crippled and the cost greatly enhanced.'[3]

This type of situation was endemic in Lancashire. The county's comparatively advanced economy, coupled with the very varied industries involved therein, meant that land was relatively scarce and there could be many competing calls on its use. Mining operations and the transportation of coal consequently did not always gain priority.

Nevertheless, coalowners exacerbated the problem. Competing against each other, trying to injure a rival, they neither co-ordinated their plans for underground operations nor acted in concert when confronted with transport problems created by other Lancashire industries. Far too often, they proved to be their own worst enemies.

This applied in another direction—drainage. While it was usual to leave barriers of coal to protect mines from inundation, owners sometimes saw no reason for doing this when they were coming to the end of their own particular workings. As a result, water would sweep through, going on to flood the mines of nearby colliery proprietors. These victims would, naturally, reply in kind should the opportunity present itself. But even where this kind of internicine warfare did not prevail, there was necessarily wasteful effort and greater expense because the coalowners would rather operate completely independent from each other, than co-operate. As the Lancashire & Cheshire Coal Association admitted: 'Each colliery has to deal with its own water, and this cannot, of course, be done at so low a cost as if several collieries were drained to a common pumping station. The difficulty is likely to be greatly increased in a few years.'[4]

The combined effect of all the previously-mentioned factors—the growing practical difficulties in mining coal, the influence of the industrial environment on the Lancashire coal industry, as well as the conduct of the coalowners themselves—were harmful. Very few new mine-shafts were sunk. Production stagnated. Costs escalated as more men were needed to maintain a given output. The introduction of new safety regulations involved extra expense and the reduction in working hours only served to aggravate the coalowners' predicament.

The following figures clearly show the dangerous trend towards diminishing returns that were present in the Lancashire and Cheshire coalfield during this period:

	Tons	*Men*
1895	22,764,171	87,227
1900	25,542,317	87,944
1910	24,627,515	104,659[5]

At about the turn of the century, Lancashire and Cheshire coalfield reached peak production. From then onwards, there was declining output.

A peculiarity of the Lancashire coalfield placed the coalowners in a vulnerable position when faced with declining production. This arose from the fact that most of the masters did not own the land in which they operated but merely leased it. The Lancashire & Cheshire Coal Association told the Royal Commission on Mining Royalties that ninety per cent of coalowners were thus placed. Most leases were for between twenty-five and fifty years, although some lasted for a hundred years or longer. Royalty was assessed by acreage, the area the coalowner had acquired the right to mine whether he did, in fact, mine it or not. Rent varied from £50 to £120 for ordinary coal and up to £200 a Cheshire acre for cannel.*

To illustrate how this system worked, let me take the case of Wigan Coal & Iron Company. In 1890, its manager, Alfred Hewlett, testified before the Royal Commission. He said that the Company's paid capital was £1,809,725, and annual output was around two million tons. The leases it acquired were 171. He divided the rent paid into three categories:

* For an explanation of the term Cheshire acre and how it was calculated see Chapter Three.

- i minimum rent, i.e. amount to be paid for the land itself £56,279
- ii overpaid on minimum, i.e. amount paid as a consequence of actual mining operations £313,642
- iii paid for coal the Company were unable to mine £60,454

Hewlett declared rents were 'a very onerous question'. They bore down heavily on colliery proprietors: 'for about ten or twelve years, there has been very little money in Lancashire, scarcely any at all'.[6]

The existing system of assessing rent was unfair, declared Hewlett, Blundell and Grinke, who testified before the Royal Commission. It took no account of the imponderables of mining. Mine disasters, industrial disputes, and numerous unforeseen things could delay the development of a coal-face and consequently retard output. As a result the coalowner was doubly penalized: loss of income through less tonnage sold and yet paying a rent for coal not mined.

Although it was not mentioned by Hewlett and his associates this system of paying rents had a further defect. Not only did the mine lessees tend to suffer disproportionately when output unexpectedly slumped but also when prices dropped. For, under these circumstances, there was, particularly with the small owners, a perverse tendency. Instead of reducing output as the market price of coal dropped, they were tempted to do the exact opposite. Aware that they would have to pay for the coal whether they mined it or not, they were liable to attempt to compensate for the drop in price by an increase in quantity. This could only create further disequilibrium between supply and demand, resulting in an even further drop in prices.

The method of rent-payment was one of the forces that helped to influence the masters' attitude towards their men. Insofar as it unduly limited profitability, it made coalowners tetchy. When this is added to other forces at work, such as the numerous geological faults which frequently altered working conditions, thus creating dispute-prone situations, and the problem of rising production costs, one can see why labour relations were so bad. Referring to Lancashire, the *Daily News* observed, 'The cumulative effect is an estrangement between masters and men such as exists nowhere else except in South Wales. For this, the nature of the coalfield even more than human beings is to blame.'[7] None of the large firms in Lancashire emerged as 'model employers', prepared to make concessions to the men and endeavour to evolve relations of amity.

Instead the Lancashire Federation of Miners was compelled to face war on two fronts: against both the small and large coalowners. While these two types of business co-existed, operated side-by-side, there were growing differences in their methods of business. Most of the small entrepreneurs had an outlook akin to that of gold prospectors. They were out to make quick profits but, alas, all too often they made quick losses. They did not possess the technique or the capital that were becoming essential as the quest for coal involved going deeper and deeper. In a forlorn struggle for survival, these small businessmen tried to extricate themselves from their predicament by attempting to cut labour costs.

The big coalowners, on the other hand, were usually content with steady, yet, substantial, returns. They did not cheesepare on the provisions for safety and ventilation to the extent of their smaller brethren. They were more concerned with continuity, maintaining and expanding production in an orderly manner. Being bigger, they tended to take a more Olympian view of the industry and made long-run calculations rather than living on a hand-to-mouth basis. Concerned about the future, their roots, in many instances, went right back to the beginnings of the industry in Lancashire.

Indeed, it was the great mining dynasties, families with interests in coal dating back to the fifteenth and sixteenth centuries, who increasingly dominated the picture as the industry entered the twentieth century. To give a few examples:

1 The House of Haigh. As stated in Chapter one, John Leland on his travels in 1538 recorded that Bradshaw was mining coal at Haigh. Since then, the business had grown immensely until it dominated the economic and political life of Wigan; the Bradshaw family had entered the nobility with a number of titles (Balcarres, Balniel, Crawford and Lindsay); and the mining interests of the family had been extended into Scotland.

2 The Knowles family. This family's interest in land and coal dated back to Elizabethan times. Andrew Knowles, born 1783, formed the colliery company of Andrew Knowles & Son which, by the end of the nineteenth century, was the largest concern in South Lancashire, employing almost four thousand workers. It was then being run by John Knowles, an alderman of the County Council, whose eldest son, Sir Lees Knowles, was M.P. for Salford, 1866-1906.

3 The Fletcher family. Their interest in the industry also dated back to Elizabethan times. By 1750, when Matthew Fletcher came to Clifton, his family owned two-thirds of the place. In 1865, the Fletchers leased all mineral rights to Evans Brothers, uncles of Pilkingtons. Edward and Alfred Pilkington were qualified mining engineers

with Evans Bros. In 1867, they created the Clifton & Kearsley Coal Company and acquired complete ownership in 1885.

4 The Bridgewater family. John Egerton, the first earl, bought the Worsley estate in 1630, when coal was already being extensively mined there. The third duke—'the canal duke'—engaged James Brindley in 1759 to construct the famous canal to Manchester that cut transport costs so considerably. He also formed, in 1803, the Bridgewater Trustees, which continued to be one of the county's big coal producers until the leases owned by the Trust expired in 1903.[8]

While the same names seemed to appear continually, giving a sense of continuity, this tended to conceal the internal changes that were taking place. Because of the amount of capital they required, coal firms were among the most eager to avail themselves of the opportunities arising from the Limited Liability Acts of 1855 and 1862. The ownership of shares was becoming divorced from the management of concerns. Indeed, growing size and complexity required a new breed of man—the mine manager who combined technical skill with economic efficiency.

Families like the Hewletts and Peases personified this trend. Almost their entire male membership, for generation after generation were trained to fill such positions. Maskell Pease was manager of Lord Crawford's mines until 1855, when he became secretary of the coalowners' national association and subsequently town clerk of Wigan. His place as manager at Haigh was taken by Alfred Hewlett who, in turn, left to be in charge of the Wigan Coal & Iron Company in 1865, a position he held for the rest of his life.

With day-to-day control in their hands, mine managers exercised immense power. They had to decide, amongst many other things, what leases should be acquired and in which direction mining operations should be directed. These decisions not merely effected their own companies' performances but also the fortunes of other coalowners and landowners in the neighbourhood.

Since, once a lease had been bought, coal had to be paid for whether it was mined or not, Lancashire mine managers were particularly favourably placed to indulge in devious practices. If they were unscrupulous, they could deliberately place money in the pockets of their friends and, indirectly, even themselves. This is precisely what Alfred Hewlett was accused of in 1870: John Lancaster, chairman of Wigan Coal & Iron Company, said that he promoted the interests of other concerns at the expense of the Company he was supposed to manage.[9] Lancaster suggested that Hewlett had been especially generous in dealings with other firms in which he (Hewlett) had shares. His former employer, Lord Crawford, received a rent from Wigan Coal & Iron Company of £17,779—£5,799 more than was fixed—while other landowners were paid £8,815 in excess rents. As Lord Crawford was a big shareholder in Wigan Coal & Iron, the Company's shareholders were divided. Both sides distributed printed booklets, attacking each other.

At the Company's extraordinary meeting, on 13 January, 1870, Hewlett disputed the accusations of mismanagement. He claimed that John Lancaster's figures were inaccurate. The actual amounts paid in excess rent were:

	Fixed	*Amount Paid in Excess*
Lord Lilford	£300	£1,414
Lord Skelmersdale	£500	£350
Miss Holcroft	£200	£695
Rev. W. Simmons	£100	£350
John Taylor	£6,000	£4,255
Lord Crawford	£12,000	£4,795

Hewlett then dwelt on the difficulties of mining. He told the extraordinary meeting that progress had been slower than anticipated and consequently development into newly acquired seams had been delayed.

Nobody appears to have been in possession of sufficiently detailed knowledge of the Company's day-to-day working to question Hewlett's claims effectively. This fact, coupled with the block vote of Lord Crawford's shares, appears to have won the day for Hewlett. Nevertheless, the suspicion still lurked in some people's minds. Twenty-three years later, the *Wigan Comet* contained an article on the Hewletts and opined that 'the Hewlett clan appropriate all the *creme de la creme* of the Wigan Coal & Iron concern, leaving nought but the three times skimmed sky-blue for the toilers and shareholders'.[10]

Many miners viewed with apprehension the gradual re-organization of the coal industry and the growing power of mine managers. Another article in the *Wigan Comet* declared that limited companies had been 'a curse to the collier'. It cited the instance of a collier who had worked for the same firm for many years. He wanted to borrow money to pay for the funeral expenses of a small child, but was refused. The paper then quoted the old saying 'a company has no soul to be saved'. A further objection was that the new-style managers often arrived at work around nine or ten o'clock. Some had 'never been down a pit in months'. Their main preoccupation, it seemed to the miners, was cutting the wage bill. The old-style managers, on the other hand, 'were not able to tell you that X equals A minus B divided by C, but they knew practically when a man did a fair day's work and whether he was earning a fair day's pay'.[11] This was partly a reference to the introduction of long wall working, where men laboured collectively on a coal-face, rather than the pillar-and-stall method, where payment depended more on individual effort. Long wall working said the *Wigan Comet*, means that 'inefficient workmen lower the pay of their fellows and bring to the employer less money for his coal through the excessive production of slack'.[12]

These criticisms show that miners, like many other kinds of workers, are conservative and do not adjust easily to new working arrangements. Their nostalgia for the past, viewing the past unhistorically and unrealistically, is based on a failure to appreciate that such changes were an inevitable concomitant of progress. It must be conceded that the long wall method made it more difficult to determine whether a particular worker was being efficient or not. Nevertheless, as a method of production it was undoubtedly superior: in large undertakings a given labour force would produce more than by the pillar-and-stall method. Similarly, their criticism of the modern manager with mathematical knowledge overlooks the fact that many of the problems facing a manager could no longer be settled by rule-of-thumb.

However much the men might have grumbled, it is reasonable to argue that these changes in the capital structure and *modus operandi* of the coal industry indirectly assisted the Lancashire Miners' Federation. First, since they were a necessity for the maintenance of existing levels of production, miners' wages and jobs depended upon them. Second, since they inevitably resulted in the decline of paternalism and the growth of non-personal management, they led to a clearer division between both sides of industry, thus creating a pre-condition for unionization of large numbers of workers. Indeed, without the development from the large coalowner into the giant corporation, the Lancashire miners probably would not have centralized their organization, a move that greatly enhanced effectiveness and bargaining power.

But the re-organization of the L.M.F. in 1897 was not in itself a panacea. Many of the most important results from centralization were not immediately felt but took a long time to mature. Consequently, at the turn of the century, Lancashire miners were still plagued with the old, perennial problems.

One such nagging problem was the treatment of those men injured or disabled in the pits. With the passing of the Employers' Liability Act 1897, the L.M.F. fondly hoped to accomplish what it had failed to do in 1881—break the shackles that bound miners to the Lancashire & Cheshire Miners' Permanent Relief Society. But employers, appreciating its value to them, resolved to stop the union. Delegates at Townley colliery reported that 'the managers had been down the pit with a book and almost coerced the men to sign their names, that they will continue to be members of Permanent Relief Society'.[13] At other places, they even made it a condition of employment. This threat was made to appear very real because, with the introduction of the new Employers' Liability Act, many coalowners made searching reviews of their own labour requirements. They did not want to continue employing men who, in any relief scheme dominated by themselves, were likely to be a constant drain on the funds. Therefore, they dismissed men suffering from 'old age, deafness, short-sightedness, or other unfortunate defects'.[14]

A spate of cases, full of human misery, came before the Lancashire Federation. In each instance, it had to adjudicate on whether the individual was entitled to assistance from the union. Members who had been victimized or made redundant were entitled; those who had little prospect of ever getting employment again were not. There were, of course, marginal cases like W. Alston of Altham. He had been working 'in a very hard place and now getting 50 years of age he cannot follow on his work every day'.[15] Then there were those who, even if they did find regular employment again, could never hope to secure a living wage. What should the Federation do, for instance, in the case of William Morecroft, of Lea Green? He lost his leg in a pit accident in 1889. After a year's convalescence, he returned to work but 'was not averaging one shilling and sixpence a day'.[16]

With such a drain on its resources arising from the membership's numerous personal problems, the Federation hardly had the time or energy to resist the coalowners' plans to compel miners to remain in the Lancashire & Cheshire Miners' Permanent Relief Society. Nevertheless, it did publish a pamphlet by Sam Woods, urging men not to contract out of the state-sponsored scheme. Twenty thousand copies were distributed. So, presumably, most miners were cognisant of the union's case against the Permanent Relief Society: the men provided seventy-nine per cent of the funds yet the owners controlled it.[17]

The Lancashire Federation also endeavoured to float its own accident and relief scheme. This was regarded either as a supplement to the state provisions or as a rival to the employer-controlled scheme. But it appears as if the project was unsuccessful. As stated in Chapter Twelve, the accident rate in Lancashire was forty per cent above that of Northumberland.[18] So, *ceteris paribus*, there would be greater calls for financial assistance. But, on the other hand, with the united opposition of the coalowners confronting them, few miners were prepared to join and consequently the resources were small. Moreover, members of the medical profession, perhaps through the incitement of the coalowners, refused to collaborate with the Federation's proposed scheme.[19]

The result was that, right through till 1947, when private ownership in the coal industry ended, Lancashire miners thought that accident and insurance provisions were inadequate and unjust. They continued to suffer under the Lancashire and Cheshire Miners' Permanent Relief Society, which remained coalowner controlled.

On another of the Lancashire Federation's problems at the turn of the century—that of the state of organization in the North-East of the county—it was able to register considerable progress. The right to appoint checkweighmen was eventually gained. But not without a big fight. In 1897, the Federation hoped to secure its goal through government intervention. At the Federation's behest, Philip Stanhope, M.P. for Burnley, raised the issue with the home secretary. He replied that he saw no reason for altering the position: exemption had been given from that particular clause of the 1872 Mines Act and he was not convinced that the majority of colliers wanted things changed.[20] The Lancashire Federation endeavoured to prove to the home secretary that he had failed to interpret the men's opinions correctly. The following year—1896—a ballot was held among the miners. By almost a three to one majority they voted for the checkweighmen. Thomas Ashton, for the Federation, informed the home secretary of the result, but this did not lead him to alter his position.

Meanwhile, amid the controversy about checkweighmen, industrial relations through the collieries in the entire North-East of the county became unsettled. Minor skirmishes were constantly occurring. In a typical month—May 1898—the Federation found itself involved in the following disputes in the Burnley area alone: the dismissal of drawers at Hapton Valley; the price for cutting at Clifton; the state of the roadways at Reedley; low coal, bad places at Burnt Hill; and cutting at Cheapside.[21]

Whenever possible, the owners discriminated against the miners who were in the union. What was known as 'rale', the distribution of workplaces among the colliers, done by minor officials done on the management's instructions, was a powerful weapon in the coalowners' hands. It was used to see that militants were situated in the worst places with the furthest to travel underground. The coalowners' favourites, who were called 'fancy men' by the rest, were permitted to use the haulage to travel to their workplaces whilst the union men were compelled to walk.[22]

From the men's standpoint, by far the most exasperating thing was the continual paring away at their wages. Shapcott, of Burnley, told a Federation conference on 20 August, 1898, 'Enchroachments were constantly being made on wages. So much so that the men are not five per cent better off than in 1888, and the men were saying what is the use of paying to a Union when these grievances cannot be remedied?'[23] This last remark was a criticism of the Federation's hesitancy, in view of painful previous experience, to become embroiled in a further all-out struggle in the Burnley area.

But really, despite its misgivings, the Federation had no alternative but to back a strike. The union membership in the North-East was restive, highly annoyed with what it regarded as the Federation's over-cautious approach. At the same time, the employers strived to widen the internal disagreements within the union's ranks. Mr R. Handsley, of the Burnley and Townley collieries, wrote in a letter to Thomas Ashton that they 'are always ready to consider and redress real complaints and perhaps in no pits do the men enjoy greater privileges or are better treated than under their firm, but we do object to grievances being manufactured and a constant state of unrest encouraged by the local committee, who seem to us anxious to thwart the Federation in what they do'.[24] For this reason, Burnley collieries declared themselves to be unwilling to negotiate with the local representatives of the union, particularly when its delegations included Sparling, but wanted to deal with the Federation directly.

This incensed the union membership all the more. Sparling told the Federation that eighteen months' of negotiations had achieved nothing. They still had the same grievances: (1) dismissal of men away from work through illness, (2) innacurate pay slips, (3) excessive deductions, (4) stopping money directly out of wage packets for the employers' accident scheme.[25] The Federation's executive eventually decided, on 7 September, 1898, to support a partial strike in the Burnley area. The men at the Townley colliery were called out and, after six months, the strike was defeated because all the jobs were taken by outsiders.[26]

While the outcome was undoubtedly a defeat for the union, there were nevertheless influences that were gradually pushing the balance of forces in its favour. First, there was the agreement with the miners' unions in other coalfields to do everything possible to discourage men from coming to North-East Lancashire when a dispute was taking place. By this period, it had become important to have a strong nucleus of experienced colliers in each pit, conditions required it; the proportion of jobs that could be done by men with no knowledge of mining was dwindling. But even among this latter group whether the supply of labourers in North-East Lancashire had grown more reluctant to work in the pits or whether, due to rural depopulation, they were not available in the same abundant numbers, it is impossible to say. But there seems to be no doubt that coalowners were finding increasing difficulty in recruiting blacklegs. Concurrently, the Lancashire Federation, thanks to its centralization, had been immeasurably strengthened, and was therefore much more able to withstand a war of attrition. So there were powerful forces influencing the coalowners of North-East Lancashire, making them see that it was in their own interests to reach some sort of working arrangement with the trade unions. In 1902, the coalowners finally agreed to checkweighing and after that date, while relations were often strained with the union, both sides of industry recognized the other side's right to exist.[27]

The Federation made a further advance insofar as, once re-organization had taken place, the tendency for disgruntled miners to leave *en bloc* ended. Localism had been tamed, not abolished. The Lancashire miners still possessed strong rivalries, were suspicious of each other, but these differences were muted and did not lead to splits in the organization. The only sphere in which the danger of serious discord existed, which could have produced serious weaknesses, came from political disagreements. In 1899, men at Bank Hall colliery, Burnley, ceded from the Burnley branch. Led by John Leeming, they stated that they objected to the politics of the Burnley branch (presumably, the socialist views of Sparling) and to the branch officers (presumably Sparling and Shapcott). Leaming was a moderate on industrial matters, with views much in line with those of the county leadership rather than the Burnley branch. He was also interested in politics. When he unsuccessfully stood for Burnley borough council in an election where Sparling was one of the socialist candidates, he was described as 'the catspaw of the Liberals'.[28] Obviously acrimony generated in the course of elections spilled over into union affairs.

But, for a short while, a more serious threat came in the form of a proposal from *The Clarion* to create an all-embracing federation of trade unionists, which would co-ordinate industrial strategy, receive quite large subscriptions and pay strike benefit.[29] Thomas Ashton and the other Lancashire miners' leaders did not regard this proposal as something that would strengthen their industrial power, providing them with fresh sources of finance, but rather as a federation that would compete with their own Federation. They also saw the danger that, through a pooling of resources, their own carefully garnered money might be squandered by less well-organized workers in other industries.

But, as we have already seen, many Lancashire miners were influenced by socialist ideas. Therefore, almost inevitably, it found able advocates in the ranks of the Lancashire Federation. Wharton Hall branch proposed that 'twelve thousand pamphlets just published by P. J. King and Robert Blatchford be purchased by this Federation and distributed amongst the members'.[30] This resolution provoked Thomas Ashton to take the unusual step of writing a personal comment: he thought the scheme 'simply impracticable and unworkable'; he hoped the Lancashire Federation would have nothing to do with it.[31]

By his action, Thomas Ashton made himself the subject of attacks from socialists, *The Clarion*, and from one of the architects of the scheme itself. P. J. King devoted quite a large portion of his pamphlet 'Good and Bad Federation' to castigating Thomas Ashton and asking:

> 'Is it not a deliberate attempt to strifle the men's opinions by prejudicing their minds so as to prevent them from reading the pamphlet and judging the scheme for themselves? Do they permit Mr. Ashton, or any other official to do all their thinking? Are the miners not capable of doing a little on their own account? Permit me also to ask Mr. Thomas Ashton, or any other of the prominent officials connected with the Federation, if he, or any of them, be willing to prove how it is impracticable? Where and how it is unworkable?'[32]

Despite opposition, Thomas Ashton succeeded in convincing the Lancashire Federation to have nothing to do with the King-Blatchford plan. It had no representatives among the 136 delegates who met at Manchester in July 1898 to launch the scheme.[33] Nevertheless, the political prescience of Ashton was vindicated by events. Even *The Clarion* was compelled to admit that the Manchester conference was attended 'by the cream if not the bulk' of the trade union movement.[34] Lack of support strangled the scheme at birth.

In conclusion, it may be worth while to summarize succinctly the main achievements of the Lancashire Miners' Federation before 1900. It had by then existed for less than twenty years. Yet it had profoundly and irrevocably altered the position of miners in the North-West coalfields. In 1881, the union membership of the entire area was, at the most, six thousand; by the turn of the century it had risen to around the forty thousand mark. Thus, two-fifths of the total labour force were organized in the Lancashire Miners' Federation. This is a greater proportion than is common in most British industries today, and was definitely greater than was prevalent in most British industries seventy years ago. When it is remembered that most of these members were underground workers, men in key positions, capable of halting production, then it is obvious that the L.M.F. was an extremely powerful organization.

Moreover, the Lancashire miners' organized strength was, to the extent that a leader controlled membership, in the hands of capable and talented officials. Leaders like Thomas Ashton had, from the Federation's inception, played an important role, had carefully husbanded the union's resources and avoided the kind of cataclysm that had doomed all previous miners' unions in the county.

Not only did the Lancashire Miners' Federation possess an experienced skilful leadership; it also provided more than its share of the men who ran the M.F.G.B. and formulated national union policy. Thomas Ashton was the first general secretary of the national body—in fact, a prime mover in its formation. He held this position from 1889 to 1919, a period much longer than any incumbent of the office has ever done since. Fellow miners' leader, William Hallam paid the following tribute to his pertinacity:

> 'Mr. Ashton's life has been spent consistently in persistent efforts to obtain justice and fair play for the miners, and his efforts had frequently resulted in loss and suffering to himself. Anyone who enjoyed the privilege of attending any of the Miners' conferences cannot fail to have observed the respect in which he is held, nor the clear and efficient manner in which he transacts the business of the meetings.'[35]

In his national history, Mr Page Arnot reaches a very similar verdict:

> 'Ashton was very methodical and persistent, careful and exact. He was no speaker, but very apt with his pen. In him the Lancashire miners first, and then the miners nationally, were to find a capable secretary and propagandist.'[36]

Whatever the limitations of Thomas Ashton's industrial and political outlook, there can be no doubt that he was more in tune with the needs of the times than his critics. Around him, the union's leadership in Lancashire grew up, stable and confident, capable of giving a lead both in the county and nationally. Sam Woods, for instance, held the position of vice-president of the M.F.G.B. longer than any other man —twenty years, 1889-1909.

With this combination of a large membership and an experienced leadership, the Lancashire Federation achieved considerable progress, reducing hours and raising pay. Of course, it was not uninterrupted progress. The business cycle still made for ups and downs. But even in the worst depression, Lancashire miners were never again reduced to the starvation levels of the winter of 1880-1. The union provided protection, a kind of safety barrier that took some of the bite out of the economic recessions.

Nevertheless, the union's powers were limited. It operated within a framework that it had not itself created. Besides the internal factors that were causing the decline of the coal industry in Lancashire, there was in the period between the First and Second World Wars another harmful influence. The Lancashire textile industry was also suffering from a severe slump, with as many as 43·2 per cent of the total labour force unemployed at the height of the depression in 1932. Obviously, since the demand for coal was a derived demand, dependent on the level of industrial activity, Lancashire coal suffered as a consequence of the reduced requirements of its main customers within the county.

Faced with considerable difficulties, the Lancashire Miners' Federation cannot always claim to have secured its objectives. It can usually, however, claim to have done as well as circumstances permitted.

Table 20 *The Clifton & Kersley Coal Co.*

	Profit & Loss
1886	£15,618
1887	£10,108
1888	£12,926
1889	£7,340
1890	£18,503
1891	£54,367
1892	£46,450
1893	£37,987
1894	£19,724
1895	£39,069
1896	£15,879
1897	£9,924
1898	£14,399
1899	£27,326
1900	£63,792

BIBLIOGRAPHY

Primary Authorities

1 *TRADE UNION RECORDS*

(a) Lancashire and Cheshire Miners' Federation: Minutes of Conferences, Executive Committee meetings, 1885-1900; balance sheets, colliery price lists, and agreements. (At the offices of the N.U.M., Miners' Hall, Bolton.)
(b) Miners' Federation of Great Britain: Minutes, circulars, reports and balance sheets from 1889 onwards.
(c) Ashton and Haydock District Miners' Association: Some reports of proceedings interspersed in bound volumes of L.M.F. minutes.
(d) Trade Union Congress: Conference reports from 1868 onwards.

2 *BUSINESS RECORDS*

(a) Astley and Tyldesley Coal and Salt Co. Balance sheets. (Lancashire Records Office.)
(b) Bridgewater Collieries Ltd. Journal-Salaries 1824-1900, Cash books 1769-1857, General Accounts 1763-1810, 1844-50. (Lancashire Records Office.)
(c) The Clifton and Kearsley Coal Co. Ltd. Stock Book, 1896-1906, Private Journal 1885-1903. Profit and Loss Balance Sheet, 1885-1922. (Lancashire Records Office.)
(d) Cliviger Coal and Coke Co. Ltd. Wage book 1854-1881. (Lancashire Records Office.)
(e) Richard Evans and Co. Ltd. Miscellaneous documents. (Lancashire Records Office.)
(f) Haigh MSS.: Accounts, correspondence, reports. (Rylands Library, Manchester.)
(g) Ince Hall Collieries. Rules of Service, Rates of Tonnage, etc., 1852-68. (Wigan Reference Library.)
(h) Tyldesley Coal Co. Reports to the Directors, 1884 onwards. (Lancashire Records Office.)
(i) Wigan Coal and Iron Co. Reports 1871-1883. (Wigan Reference Library.)

3 *MISCELLANEOUS MANUSCRIPTS, LIBRARY COLLECTIONS, ETC.*

(a) Bell and Buddle Collections, containing material relating to the Miners' Association. (North of England Mining Engineers' Library, Newcastle-on-Tyne.)
(b) The History of the Coal Trade, 12 vols. (Picton Library, Liverpool.)

(c) Holt-Gregson Papers. (Liverpool Records Office.)
(d) Hulton of Hulton. Miscellaneous Papers: Labour Disturbances, 1819-30. (Lancashire Records Office.)
(e) Howell Collection. (Bishopgate Institute, London.)
(f) Lancashire Records Office: twelve bundles of unsorted documents relating to the coal industry.
(g) Correspondence of E. L. Stanley. (Ryland Library, Manchester.)
(h) Swinton and Pendlebury Reference Library Mining Collection.
(i) Wigan Mining and Technical College: unsorted collection on the coal industry.
(j) Wigan Reference Library: Wigan Broadsides (3 vols), Pitman's Strike Collection; Pamphlets relating to Mining and Manufacture (20 vols).
(k) Bradshaw-Gerrard Mining Dispute, 1634-6.

4 *PARLIAMENTARY PAPERS AND OTHER OFFICIAL PAPERS*

Children's Employment Commission, 1842.
Reports of H.M. Inspectors of Mines and Quarries, 1844-1900.
Select Committee on Coal Mines, 1852.
Select Committee on Accidents, 1854.
Report of Commission on the State of Popular Education in England, 1861.
Select Committee on Mines, 1866.
Royal Commission to Enquire into Trade Unions and other Associations, 1867-1869.
Royal Commission on Coal, 1871.
Select Committee on the Scarcity and Dearness of Coal, 1873.
Royal Commission on the Labour Laws, 1875.
Royal Commission on the Depression of Trade, 1886.
Royal Commission on Mining Royalties, 1890-1.
Royal Commission on Labour, 1892.
Hansard reports of Parliamentary Debates.
Home Office Papers.

5 *BOOKS ON MINING, MOSTLY OF A PRIMARY NATURE*

ARNOT, ROBIN PAGE, *The Miners*. Allen & Unwin, London, 1949.
—— *The Miners: Years of Struggle*. Allen & Unwin, London, 1953.
—— *A History of the Scottish Miners*. Allen & Unwin, London, 1955.
—— *South Wales Miners*. Allen & Unwin, London, 1967.
ASHTON, THOMAS, *Three Big Strikes in the Coal Industry*, 3 vols. M.F.G.B., Manchester, 1925.
ASHTON, T. S. and SYKES, *The Coal Industry in the Eighteenth Century*. Manchester University Press, 1925.
BOYD, R. N., *Coal Pits and Pitmen*. Whittaker and Co., London, 1892.
—— *Mines Inspection*. Whittaker and Co., London, 1871.
BUNNING, T. WOOD, *An Account of the Duties of Coal*. Andrew Reid, Newcastle-on-Tyne, 1883.
CHALLINOR, R. and RIPLEY, B., *The Miners' Association: A Trade Union in the Age of the Chartists*. Lawrence and Wishart, London, 1968.

DALZIEL, A., The Colliers' Strike in South Wales. *Western Mail*, Cardiff 1872.

DUNN, M., *Treatise on the Winning and Working of Collieries*. Published by the author, Newcastle-on-Tyne, 1852.

EDWARDS, NESS, *History of the South Wales Miners' Federation*. Lawrence and Wishart, London, 1938.

EVANS, E. W., *The Miners of South Wales*. University of Wales, Cardiff 1961.

FYNES, RICHARD, *The Miners of Northumberland and Durham*. Thomas Summerbell, Sunderland, 1873.

GIBSON, WALCOTT, *Coal in Great Britain*. Arnold, London, 1927 ed.

GALLOWAY, R. L., *Annals of Coal Mining and the Coal Trade*, 2 vols Colliery Guardian Co., London, 1907.

GREGORY, ROY, *The Miners and British Politics, 1900-1914*. Oxford University Press, 1968.

JEVONS, H. S., *The British Coal Trade*. Kegan Paul, London, 1915.

LAIRD, W., *Export Coal Trade of Liverpool*. Johnson, Liverpool, 1850.

MACHIN, FRANK, *The Yorkshire Miners*. Yorkshire N.U.M., Barnsley, 1954.

MORRIS, J. H. and WILLIAMS, L. T., *The South Wales Coal Industry*. University of Wales, Cardiff, 1958.

NEF, J. U., *The Rise of the British Coal Industry*, 2 vols. Routledge, London, 1932.

PERCY, C. M., *Mining in the Victorian Era*. Wall, Wigan, 1897.

RAYNES, JOHN R., *Coal and its Conflicts*. Benn, London, 1928.

ROWE, J. W. F., *Wages in the Coal Industry*. P. S. King and Co., London, 1923.

WEBB, SIDNEY, *Story of the Durham Miners (1662-1921)*. Fabian Society, London, 1921.

WELBOURNE, E., *The Miners' Unions of Northumberland and Durham*. Cambridge University Press, 1923.

WILLIAMS, D. J., *Capitalist Combination in the Coal Industry*. Labour Publishing Co., London, 1924.

WILLIAMS, J. E., *The Derbyshire Miners*. Allen & Unwin, London, 1962.

WILSON, J., *History of the Durham Miners' Association*. J. H. Veitch, Durham, 1907.

6 *ARTICLES IN SPECIALIST PERIODICALS*

ALLEN, V. L., The Origins of Industrial Conciliation and Arbitration. *International Review of Social History*, vol. ix (1964).

ANDERSON, D., Blundell's Collieries: Wages, Disputes and Conditions of Work. *Transactions of the Historic Society of Lancashire and Cheshire*, vol. 117, 1965.

—— Blundell's Collieries: Technical Developments, 1776-1966. *Transactions of the Historic Society of Lancashire and Cheshire*, vol. 119, 1967.

BANKES, J. H. M., The Nineteenth-Century Colliery Railway. *Transactions of the Historic Society of Lancashire and Cheshire*, vol. 114, 1962.

CLEGG, H. E., Historical Notes on the Wigan Coalfield. Manchester Geological and Mining Society, 1957.

CLEMENTS, R. V., British Trade Unions and Popular Political Economy, 1850-1875. *Economic History Review*, vol. xiv, No. 1, August 1961.

COLE, G. D. H., Some Notes on British Trade Unionism, in *Essays in Economic History*, vol. iii, (ed. E. M. Carus-Wilson). Edward Arnold, London, 1962.

CRAWFORD, EARL OF, Haigh Cannel. Manchester Statistical Society, 1933.

CROFTON, H. T., Lancashire and Cheshire Coal Mining Records. *Lancashire and Cheshire Antiquarian Society*, vol. vii, 1889.

DICKINSON, J., Statistics of Collieries of Lancashire, Cheshire and North Wales. *Memoirs of the Manchester Literary and Philosophical Society*, 1855.

HAIR, P. E. H., Mortality from Violence in British Coal Mines, 1800-1850. *Economic History Review*, new series, vol. 21, December 1968.

HARRIS, J. P., The Hughes Papers: Lancashire Social Life, 1780-1825. *Transactions of the Historic Society of Lancashire and Cheshire*, vol. 103, 1951.

HASSALL, E. R. and TRICKERT, J. P., The Duke of Bridgewater's Underground Canals. *The Mining Engineer*, October 1963.

MACKENZIE, R. T. and SILVER, ALAN, Conservatism, Industrialism and the Working Class Tory of England (a paper presented to 5th World Congress of the International Sociological Association, 1962).

MCCORMICK, B. and WILLIAMS, J. E., Miners and the Eight-Hour Day. *Economic History Review*, 1959.

MACKWORTH, HERBERT, On the Diseases of Miners. *Journal of the Society of Arts*, April 1855.

MARSHALL, JOHN D., The Lancashire Rural Labourer. *Transactions of the Lancashire and Cheshire Antiquarian Society*, 1961.

ROSE, A. J., Early Cotton Riots in Lancashire, 1769-1779. *The Transactions of the Lancashire and Cheshire Antiquarian Society*, vol. 73-4, 1963-4.

SMETHURST, J. B., Strikes and Strikebreakers in the Worsley Coalfield. Eccles Historical Society, 1968.

—— Lancashire and the Miners' Association of Great Britain and Ireland. Eccles Historical Society, 1969.

TAYLOR, A. J., The Miners' Association of Great Britain and Ireland. *Economica*, February 1955.

—— Productivity and Technological Innovation in the British Coal Industry, 1850-1914. *Economic History Review* (2nd series) vol. xiv, August 1961.

—— The Sub-Contracting System in the British Coal Industry, in *Studies in the Industrial Revolution* (ed. L. S. Pressnell).

—— The Wigan Coalfield in 1851. *Transactions of the Historic Society of Lancashire and Cheshire*, vol. 106, 1954.

VINCENT, JOHN, Second Reform Act in Lancashire. *The Historical Journal*, vol. 11, 1968.

WHITAKER, J. and WILKINSON, T. T., Abstracts of the Principal Mines of the Burnley Coalfield. *Transactions of the Historic Society of Lancashire and Cheshire*, vol. ii, 1861-2.

YOUNGSON-BROWNE, A. J., Trade Union Policy in the Scots Coalfield, 1855-1885. *Economic History Review* (2nd series), vol. vi, April 1954.

7 *NEWSPAPERS AND JOURNALS*

I LOCAL: *Bolton Chronicle, Bolton Free Press, Bolton Evening News, Bristol Times, Kilmarnock Times, Leigh Chronicle, Liverpool Mercury, Liverpool Weekly Albion, Manchester Courier, Manchester Examiner, Manchester Mercury, Merthyr Tydvil Express, Oldham Examiner and Times, Preston Chronicle, Preston Pilot, Sheffield Telegraph, Swinton and Pendlebury Journal, South Wales Daily News, Wakefield Journal, Western Mail, Wigan Comet, Wigan Examiner, Wigan Gazette, Wigan Observer, Wrexham Advertiser.*

II NATIONAL: *Daily Mail, Manchester Guardian, Daily News, The Times.*

III MISCELLANEOUS JOURNALS: *The Contemporary Review, London Illustrated News, National Reformer, New Review, Nineteenth Century, North of England Magazine, Pall Mall Gazette, Practical Magazine, The Spectator.*

IV RADICAL/UNION/SOCIALIST JOURNALS: *The Beehive, Bolton Star, Burnley Socialist and North-East Lancashire News, The Clarion, Co-operative News, Flint Glass Makers' Magazine, Friend of the People, Justice, Northern Star, Labour Elector, Labour Leader, Labour News, Labour Press and Workmen's Examiner, Pendlebury Pioneer, The People's Paper, Red Republican, United Trades Co-operative Journal, The Voice of the People, Workmen's Times.*

V MINING JOURNALS: *The British Miner, Colliery Guardian, Lancashire Miners' Federation Journal, The Miner, Miners' Advocate, Miners' and Workmen's Advocate, Miners' Journal, Miners' Magazine, The Mining Engineer, The Proceedings of the North of England Mining Engineers.*

8 *BOOKS ON LABOUR HISTORY*

ASPINALL, A., *The Early Trade Union.* Batsford, London, 1949.

BEALEY, F. and PELLING, H., *Labour and Politics, 1900-1906.* Macmillan, London, 1963.

BRIGGS, ASA and SAVILLE, JOHN, *Essays in Labour History.* MacMillan, London, 1960.

CLEGG, H. A., FOX, A. and THOMPSON, A. F., *A History of British Trade Unions since 1889.* Oxford University Press, 1964.

COLE, G. D. H., *A Short History of the British Working Class Movement.* Allen & Unwin, London, 1948.

—— *Attempts at a General Union, 1818-1834.* MacMillan, London, 1953.

—— *British Working Class Politics, 1832-1914.* Routledge, London, 1941.

COLE, G. D. H. and POSTGATE, RAYMOND, *The Common People.* Methuen, London, 1949 ed.

COMPTE DE PARIS, *Trade Unions of England.* Smith, Elder and Co., London, 1869.

DOUSE, R. E., *Left in the Centre.* Longmans, London, 1966.

GILLESPIE, F. E., *Labour and Politics in England, 1850-1867.* Cass, London, 1968.

HAMMOND, J. L. and B., *The Town Labourer.* Longmans, London, 1917.

—— *The Skilled Labourer.* Longmans, London, 1909.

Harrison, Royden, *Before the Socialists*. Routledge & Kegan Paul, London, 1965.
Hobsbawm, Eric, *Labouring Men*. Weidenfeld and Nicolson, London, 1964.
Howell, George, *Labour Legislation, Labour Movements and Labour Leaders*, 2 vols. T. Fisher Unwin, London, 1902.
Humphrey, A. W., *A History of Labour Representation*. Constable, London, 1912.
Joll, James, *The Second International, 1889-1914*. Weidenfeld and Nicolson, London, 1968.
Pelling, Henry, *A History of British Trade Unionism*. MacMillan, London, 1963.
—— *Origins of the Labour Party*. Oxford University Press, 1965.
Poirer, Philip, *The Advent of the Labour Party*. Allen & Unwin, London, 1958.
Robert, B. C., *The Trade Union Congress, 1868-1921*. Allen & Unwin, London, 1958.
Saville, John (ed.), *Democracy and the Labour Movement*. Lawrence and Wishart, London, 1954.
Thompson, E. P., *The Making of the English Working Class*. Gollancz, London, 1963.
Turner, H. A., *Trade Union Growth, Structure and Policy*. Allen & Unwin, London, 1962.
Wearmouth, R. F., *Methodism and the Struggle of the Working Classes, 1850-1900*. Edgar and Backus, London, 1954.
—— *Some Working Class Movements of the Nineteenth Century*. Epworth Press, London, 1948.
Webb, S. and B., *Industrial Democracy*. Longmans, London, 1911.
—— *History of Trade Unions, 1666-1920*. Longmans, London, 1920 ed.

9 *AUTOBIOGRAPHIES, BIOGRAPHIES AND MEMOIRS*

Armytage, W. H. G., *A. J. Mundella, 1825-1897. The Liberal Background to the Labour Movement*. Benn, London, 1951.
Authors various, *Champion of Liberty: Charles Bradlaugh*. London, 1933.
Bamford, Samuel, *Life of a Radical and Early Days*. T. Fisher and Unwin, London, 1893.
Bent, James, *Criminal Life: Reminiscences of Forty-two years a Police Officer*. Heywood, Manchester, 1891.
Bonner, Hypatia, *Charles Bradlaugh. A Record of His Work*, 2 vols. T. Fisher and Unwin, London, 1902.
Cooper, Thomas, *The Life of Thomas Cooper*. Hodder and Stoughton, London, 1872.
Evans, E. W., *Mabon. A Study in Trade Union Leadership*. University of Wales, 1959.
Hallam, William, *Miners' Leaders*. Thirty Portraits—Biographical Sketches. Bemrose and Sons, London, 1894.
Hulbert, H. L. P., *Sir F. S. Powell, Bart and M.P.; A Memoir by his nephew*. Richard Johnson, Leeds, 1914.
Monypenny, W. F. and Buckle, G. E., *The Life of Benjamin Disraeli*, vol. iv. John Murray, London, 1910-20.

SHEENY-SKEFFINGTON, F., *Michael Davitt.* MacGibbon and Kee, London, 1967.
STEWART, WILLIE, *Keir Hardie.* I.L.P., London, 1925.
THOMPSON, LAWRENCE, *Robert Blatchford.* n.p., n.d.
TORR, DONA, *Tom Mann and His Times.* Lawrence and Wishart, London, 1956.
TSUZUKI, C., *H. M. Hyndman and British Socialism.* Oxford University Press, 1961.
WATSON, AARON, *T. Burt: A Great Labour Leader.* Brown, Langham and Co., London, 1908.
WEBB, BEATRICE, *Diaries.* Longmans, London, 1926.

10 *PAMPHLETS, BOOKLETS ETC.*

ANON., An Order for the Pit, a description of Fletcher, Burrows and Co., Atherton. E. Hulton, Manchester, n.d.
ASHWORTH, H., *The Preston Strike.* An Inquiry into its Causes and Consequence. Published by the author, Bolton, 1854.
AXON, W. E. A., *The Miners' Doom.* Heywood, Manchester, 1885.
BAINBRIDGE, E., The Coal Strike of 1893. *Colliery Guardian,* London, 1894.
BURNLEY MINERS' ASSOCIATION, *Full Report of The Great Conspiracy Trial of 1874.* W. Smith, Burnley, 1874.
CAMPBELL, GEORGE, L., *Miners' Insurance—their origin and extent.* Waterlow and Son, London, 1885.
—— *Miners' Thrift and Employers' Liability.* Strowger and Son, Wigan, 1891.
—— National Relief caused by accidents in Mines. *Wigan Observer,* 1878.
—— *The Permanent Fund and the Compensation Act.* Bayley and Bradley, Wrexham, 1898.
CLAPP, W. J., *Improved Machinery and Apparatus.* Nant-y-glo, n.d.
COLWELL, C., *Fiery Facts.* Effingham Wilson, London, 1861.
DAVIDSON, J. M., *Mining Rents and Royalties.* Pamphlets for the People, S.D.F., n.d.
DUNN, M., Accidents in Coal Mines. *Daily Journal,* Newcastle-on-Tyne, 1854.
GWYN, W. B., *Democracy and the Cost of Politics.* London University Press, 1962.
HANHAM, H. J., *Elections and Party Management.* Longmans, London, 1959.
JONES, BENJAMIN, *Co-operative Production.* Oxford, 1894.
KAY, M., *The History of Rivington and Blackrod Grammar School.* Manchester University Press, 1931.
LLOYD, TREVOR, *The General Election of 1880.* Oxford University Press, 1968.
McKENZIE, R. T., *British Political Parties.* Heinemann, London, 1955.
McKENZIE, R. T. and SILVER, ALAN, *Angels in Marble.* Heinemann, London, 1968.
MARX, KARL, *Capital,* vols 3. Foreign Languages Publishing House, Moscow, 1962.
MARX-ENGELS on Britain, Foreign Languages Publishing House, Moscow, 1955.

MIDWINTER, E. C., *Victorian Social Reform.* Longmans, London, 1969.
MOFFIT, L. W., *England on the Eve of the Industrial Revolution.* Cass, London, 1958.
NORDLINGER, ERIC, *The Working Class Tories.*
O'LEARY, C., *The Elimination of Corrupt Practices in British Elections, 1868-1911.* Oxford University Press, 1962.
PELLING, HENRY, *Social Geography of British Elections, 1885-1910.* MacMillan, London, 1967.
PERKIN, HAROLD, *The Origins of Modern English Society, 1780-1880.* Routledge & Kegan Paul, London, 1969.
PHILLIPS, J., *General History of Inland Navigation.* Evans, London, 1794.
PLUMP, J. H., *Men and Places.* Penguin, London, 1967.
REDFORD, A., *Labour Migration in England, 1800-1850.* Manchester University Press, 1964.
RUSKIN, JOHN, *Fors Clavigeria.* Allen & Unwin, London, 1908.
SAUNDERS, W., *The New Parliament, 1880.* Cassell, Petter, Galpin and Co., London, n.d.
TYLECOTE, MURIEL, *The Mechanics' Institutes of Lancashire and Yorkshire before 1851.* Manchester University Press, 1957.
WINSTANLEY, GERRARD, *Works of Winstanley* (ed. L. Hamilton). Cresset Press, London, 1944.
DUNN, M., *A History of the Steam Jet.* Published by author, Newcastle-on-Tyne, 1854.
FAIRLEY, W., *Calculations on the Produce of Coal Seams.* W. M. Hutchings, London, 1867.
HEWLETT, W., *The Wigan Coal and Iron Company.* A Reply by Mr Hewlett. Spottiswoode and Co., London, 1870.
HOWCROFT, JOHN, SEN., Address to the Coal Miners of Ashton-under-Lyne and its Vicinity. T. Ashton, Openshaw, 1897.
KING, P. J., Good and Bad Federations. *The Clarion*, Manchester, 1897.
MACDONALD, ALEXANDER, Address to Miners of the United Kingdom. *Glasgow Sentinel*, 1878.
—— The Mines Regulation Act. *Glasgow Sentinel*, 1872.
MUSSON, A. E., *The Congress of 1882.* T.U.C., London, 1955.
PHILLIPS, J., *General History of Inland Navigation.* Evans, London, 1794.
PERCY, C. M., *Mining in the Victorian Era.* Wall and Co., Wigan, 1878.
ROBY, GEORGE, Lamp 63—the Abram Colliery Explosion. *The Guardian*, Bolton, 1884.
SOUTH WALES MINERS, *The Double and Single Shift Compared.* Merthyr, 1868.
TUFNELL, E. C., *The Character, Object and Effects of Trade Unions.* London, 1834. (A pamphlet in the British Museum.)
WYNTER, ANDREW, *My Lords of Coal.* n.p., n.d.

11 *LOCAL HISTORIES*

AIKEN, J., *A description of the Country from thirty to forty miles around Manchester.* Heywood, Manchester, 1897.
BAINES, E., *History of the County Palatine and the Duchy of Lancaster.* Fisher, London, 1836.

BARKER, T. C. and HARRIS, J. R., *A Merseyside Town in the Industrial Revolution—St. Helens, 1750-1900.* Liverpool University Press, 1954.

BENNETT, W., *History of Burnley.* Burnley Corporation, 1951.

—— *History of Marsden and Nelson.*

BROWN, W. HENRY, *Wigan Welfare.* Wigan and District Equitable Co-operative Society. n.d.

HARDWICK, C., *History of the Borough of Preston and its Environs.* Worthington and Co., Preston, 1857.

HEWITSON, A., History of Preston. *Preston Chronicle,* 1883.

MIDWINTER, E. C., *Law and Order in Early Victorian Lancashire.* York University, 1968.

LUNN, J., *Leigh: The Historical Past of a Lancashire Borough.* Leigh Corporation, 1960.

PICTON, *Memorials of Liverpool.* Longman, Green and Co., London, 1873.

Victoria History of the County of Lancashire. Constable, London, 1908.

12 *SECONDARY AUTHORITIES*

ARNOLD, MATTHEW, *Irish Essays.* Smith, Elder and Co., London, 1891.

BARKER, T. C., *Pilkington Brothers and the Glass Industry.* Allen & Unwin, London, 1960.

BARRY, E. ELDON, *Nationalisation in British Politics.* Jonathan Cape, London, 1965.

BOER, SAMUEL H., *Modern British Politics.* Faber, London, 1965.

BLAKELEY, BERNARD, *A History of St. Stephen's School, Kearsley.* Published privately, n.d.

BOWLEY, A. J., *Wages in the United Kingdom.* Cambridge, 1937.

BRADLAUGH, CHARLES, *Humanity's Gain from Unbelief.* Watts and Co., London, 1929.

CARUS-WILSON, E. M., *Essays in Economic History,* vol. iii.

COLE, G. D. H., *A Century of Co-operation.* Allen & Unwin, London, 1944.

CUBBON, WILLIAM, *Bibliography of Works relating to the Isle of Man.* Douglas, 1934.

DODD, A. H., *The Industrial Revolution in North Wales.* University of Wales Press, Cardiff, 1933.

DODD, MAURICE, *On Economic Theory and Socialism.* Routledge & Kegan Paul, London, 1955.

DODD, WILLIAM, *Factory System.* Stoughton, London, 1842.

ENGELS, FREDERICK, *Conditions of the English Working Class in 1844.* Allen & Unwin, London, 1948 ed.

FORSTER, JOHN MONK, Children of Darkness. *The Comet,* Wigan, 1893.

—— A Pit-Brow Lassie, *The Comet,* Wigan, 1894.

NOTES ON CHAPTER 1

1 *Whittaker's History of Manchester*, cited by H. T. Crofton, Lancashire and Cheshire Mining Records, p. 32.

2 A. Gaskell, *Our Rude Forefathers*, p. 2. (Typescript in Swinton and Pendlebury Public Library.)

3 J. U. Nef, *The Rise of the British Coal Industry*, Routledge, London vol. i, p. 6.

4 R. L. Galloway, *Annals of Coal Mining and the Coal Trade*, Colliery Guardian Co. London (1907) pp. 28-29; J. U. Nef, op. cit., vol. i, p. 8.

5 J. U. Nef, vol. i, p. 64.

6 J. Leland, *Itinerary*, vol. vii, pp. 47-8.

7 Earl of Crawford, *Haigh Cannel*, p. 8, Manchester Statistical Society, 1933.

8 *Parliamentary Survey*, Lancashire, nos. 8 and 19.

9 J. U. Nef, op. cit., vol. i, p. 322.

10 *Duchy of Lancaster Decrees and Orders*, vol. 28, f.210b, cited H. T. Crofton, op. cit., pp. 57-8.

11 *Duchy of Lancaster Pleadings*, 138/L/13.

12 *Palatine of Lancashire Bills*, 38/27, 39/53.

13 J. U. Nef, op. cit., p. 353.

14 *Haigh MSS.* 12 August, 1688. (Rylands Library, Manchester.)

15 *Haigh MSS.* show that Sir Roger Bradshaw built the Great Haigh Sough for posterity and he succeeded. It was only in 1932, when it fell into disuse, that extensive flooding took place in the Aspull-Westhoughton area. A law suit, said to have cost £50,000, ensued. Sir Stafford Cripps and Lord Shawcross, both barristers, appeared for the contending coalowners.

16 Cannel is high quality coal. Winstanley's report indicates that the practice of giving miners concessionary coal dates back at least into the seventeenth century. Another Winstanley—Gerrard Winstanley, the Diggers' leader—who also came from Wigan but appears not to have been related to his namesake, referred to cannel in his pamphlet, Law of Freedom; True Magistracy (1651), and talks about the importance of mining as a source of employment. See *Selections from the Works of Gerrard Winstanley* (ed. L. Hamilton), The Crescent Press, London (1944) p. 175.

17 H. E. Clegg, *Historical Notes on the Wigan Coalfield*, p. 8, Manchester Geological and Mining Society.

18 *Mining dispute*, 1634-6, a document in Wigan Public Library.

19 Picton, *Memorials of Liverpool*, vol. i, p. 131. The difficulty of access resulted in Liverpool having, during the eighteenth century, almost a permanent shortage of coal. Dr J. R. Harris writes, 'The preoccupation of Liverpool's citizens with this coal question makes us realise how serious Merseyside's persistent fuel crisis seems to contemporaries, and it partly explains why waterways so advantageous to the port as the Trent and Mersey and the Bridgewater canals had proportionately so little space devoted to them in the controversies of the time'. *Liverpool and Merseyside*, a book of essays ed. by J. R. Harris, p. 94.

See Table 2 for the great increase in coal entering Liverpool after these improvements in communications.

20 Arthur Young, *A Six Months Tour through the North of England*, quoted by E. Lipson, The Growth of English Society, p. 228.

21 E. R. Hassall and J. P. Trickett, *The Duke of Bridgewater's Underground Canals, The Mining Engineer*, October 1963.

22 J. Phillips, *General History of Inland Navigation*, p. 76.

23 At Harecastle, on the Cheshire-Staffordshire border, where the canal cuts through the hill, bargemen had to lie on their backs and kick the tunnel's roof to push their vessels along.

24 J. Aiken, *A description of the Country from thirty to forty miles round Manchester*, p. 115; J. Phillips, op. cit., p. 76.

25 Quoted by J. H. Plumb in his essay, 'Josiah Wedgwood', published in his book, *Men and Places*, p. 133.

26 In 1808, Balcarres bought Gerrard's 208 acres for less than £80 an acre. The Earl of Crawford said, 'This was a lucrative enterprise. On the other hand, he amalgamated a large group of scattered mineral royalties, which proved worthless, and after a barren century technically stands in my books as a debit of £100,000.' Earl of Crawford, *Haigh Cannel*, p. 19.

27 L. W. Moffit, *England on the Eve of the Industrial Revolution: A Study of Economic and Social Conditions from 1740 to 1760, with special reference to Lancashire*, p. 268.

28 See Bridgewater Colliers Ltd, company books, Lancashire Public Records Office.

29 A. J. Taylor, 'The Sub-Contract System in the British Coal Industry', *passim*, an essay in *Studies in the Industrial Revolution* (edited by L. S. Pressnell).

30 H. T. Crofton, op. cit., p. 61.

31 J. U. Nef, op. cit., vol. ii, p. 156.

32 *Haigh MSS.* William Porter's Day Book, 1744-55.

33 T. S. Ashton and J. Sykes, *The Coal Industry in the Eighteenth Century*, p. 99.

34 Hylton Scott, *The History of the Miners' Bond in Northumberland and Durham*, a typescript in Newcastle-on-Tyne Public Library.

35 R. Challiner and B. Ripley, *The Miners' Association: A Trade Union in the Age of the Chartists*, chapter 5: Men were expected to work at Wingate after the winding-gear had broken; at Seaton Delaval when the pit was still on fire; and at Thornley when fines exceeded wages.

36 The Master and Servants Acts gave employers similar powers until 1867, if not till 1875. Many thousands of workers were punished under this legislation. However, coalowners did not turn to the Master and Servant Acts with the same readiness as they did to the bond. It may be that this was because the bond was regarded as something personal, arrived at by mutual arrangement between master and men, whereas the Master and Servant Acts seemed to be, as it were, imposed from without. When masters thought men had broken their bond, this was taken as a personal affront that could not be overlooked. The bond, therefore, was regarded by miners as a greater menace than the Master and Servant Acts, a point that is clearly shown by the struggles of colliers in the North-East coalfields to get it abolished.

37 *Manchester Mercury*, 13 June, 1757; J. B. Smethurst, Strikes and Strikebreakers in the Worsley Coalfield, p. 1, a pamphlet published by Eccles and District History Society in 1968.

38 D. Anderson, Blundell's Collieries: Wages, Disputes and Conditions of Work, p. 109, in *Transactions of the Historic Society of Lancashire and Cheshire*, vol. 177, 1965. Also, A. G. Rose, Early Cotton Riots in Lancashire, 1769-1779, *passim*, in *The Transactions of the Lancashire and Cheshire Antiquarian Society*, vols. 73-4, 1963-4.

39 Home Office paper 42/20, cited by A. Aspinall, *The Early English Trade Unions*, p. 3.

40 Home Office paper 42/22, cited by A. Aspinall, op. cit., p. 7.

41 Eric Hobsbawn, 'The Machine Breakers', an essay in the book, *Labouring Men*, p. 16.

42 G. Derbyshire, *A Short History of Trade Unionism*, p. 15, a thesis in Wigan Public Library. Also, D. Anderson, op. cit., p. 115.

43 During the trial of fifteen workers at Manchester for conspiracy in 1819, it was stated that, 'All societies, whether benefit societies or otherwise, were only cloaks for the people of England to conspire against the State'. P. H. J. H. Gosden, in *The Friendly Societies of England*, 1815-1875, p. 71, points out that trade unions often operated under the guise of being simply friendly societies.

44 The textile workers asked the miners to join them in a General Union according to J. L. and B. Hammond. See *The Skilled Labourer*, pp. 103-4.

45 R. F. Wearmouth, *Some Working Class Movement of the Nineteenth Century*, pp. 300-1.

46 Home Office paper 42/179, cited by A. Aspinall, op. cit., p. 279. Joseph Healey was a Bolton weaver who became a quack doctor. In 1817, he was arrested with Samuel Bamford. Two years later, he was one of the organizers of the Manchester meeting at St Peter's Fields, and subsequently was sentenced to a year's imprisonment for sedition. See Samuel Bamford, *Life of a Radical and Early Days*, vol. ii, pp. 45-53, 251.

47 Home Office papers, cited A. Aspinall, op. cit., p. 293.

48 Home Office papers, H.O. 42/180, cited A. Aspinall, op. cit., p. 288.

49 In his book, *The Making of the English Working Class*, E. P. Thompson opines, 'From the Public Records Office, Bolton appears to have been the most insurrectionary centre in England from the late 1790s until 1820. But it is by no means clear whether this was because Boltonians were exceptionally revolutionary in disposition; or, because Bolton suffered from two unusually zealous magistrates—the Rev. Thomas Bancroft and Colonel Fletcher—both of whom employed spies (or "missionaries") on an exceptional scale.' (pp. 490-1.)

50 Home Office paper 42/180, cited A. Aspinall, op. cit., p. 302.

51 Home Office papers 41/4/253-4, cited A. Aspinall, op. cit., pp. 317-8.

52 *Liverpool Mercury*, 19 February, 1819.

53 Home Office paper 42/185, cited A. Aspinall, op. cit., p. 318.

54 J. R. Harris, *The Hughes Papers: Lancashire Social Life*, 1780-1825, p. 125, in *Transactions of the Historic Society of Lancashire and Cheshire*, vol. 103, 1951. This article contains an extremely full and detailed account of the strike.

55 The first use of the Lancashire strike tactic is usually ascribed to John

Doherty in the cotton spinners' strike of 1830. (See S. & B. Webb, *The History of Trade Unionism*, pp. 119 and 122.) The St Helens miners appear to have anticipated him by eleven years.

56 E. W. Evans, *Miners of South Wales*, p. 42.

57 J. E. Williams, *The Derbyshire Miners*: 'thousands of colliers in Derbyshire were persuaded by delegates from Lancashire to join Doherty's National Association for the Protection of Labour in 1831'. (p. 88.)

58 Richard Fynes, *The Miners of Northumberland and Durham*, pp. 19-32.

59 *United Trades Co-operative Journal*, 27 March, 1830. Also, G. D. H. Cole, *Attempts at a General Union*, pp. 32-3.

60 G. D. H. Cole, op. cit., p. 35.

61 *The Voice of the People*, 30 April, 1831.

62 D.D.H.U./53/72, Lancashire Records Office.

63 *The Voice of the People*, 28 May, 1831. The mine disaster is also mentioned in Notes from an Old Clifton Diary in Swinton and Pendlebury Public Library.

64 *The Voice of the People*, 18 June, 1831.

NOTES ON CHAPTER TWO

1 Three histories of the Miners' Association have so far been written: A. J. Taylor's article, 'The Miners' Association of Great Britain and Ireland', *Economics*, February 1955; R. Challinor and B. Ripley, '*The Miners' Association: A Trade Union in the Age of the Chartists*'; and John Smethurst's regional study, '*Lancashire and the Miners' Association of Great Britain and Ireland*', Eccles Historical Society 1969.

2 *Northern Star*, 29 May, 1847.

3 David Swallow, John Berry and others who were subsequently to become members of the Miners' Association testified before the Children Employment Commission. The union supported the Ashley Act of 1842, although it was unpopular with some colliers because, with women and children not working underground, the family incomes were reduced.

4 S. and B. Webb, *The History of Trade Unionism*, p. 183; R. Challinor and B. Ripley, op. cit., p. 214.

5 W. P. Roberts (1806-71), youngest son of the Rev. Thomas Roberts, educated at Charterhouse. In 1828, admitted as a solicitor, practised at Bath, Newcastle-on-Tyne and Manchester. In 1838, met Henry Vincent, became a Chartist, and was arrested at Trowbridge the following year. From 1843 onwards, worked for the Miners' Association and other trade unions. In 1844, legal adviser to the Chartist Land Bank, in which scheme he lost considerable money. In 1862-3, visited the Holy Land, and afterwards lectured on Biblical subjects. In 1867, assisted in the defence of Allen, Larkin and O'Brien ('the Manchester martyrs'), hanged for killing a policeman. Buried at Chorley Wood, Rickmansworth.

(Sources: *Dictionary of National Biography* and C. W. Sutton's biographical note in the Manchester Reference Library.)

6 *Bolton Chronicle*, 31 January, 1846. The remarks were made by William

Dixon when he chaired the public meeting at the beginning of the Bolton strike.

7 *Flint Glass Makers' Magazine*, October 1851.

8 *Miners' Advocate*, 24 August, 1844: 'There are now upwards of 100,000 men in the Union and we are increasing from day to day.' S. and B. Webb, op. cit., p. 182, give the figure of 100,000.

9 *Northern Star*, 27 January, 1849. The same article makes the point 'there was no class of labourers who stood in a better position to resist the aggressions of the strongest class of the masters'.

10 *Miners' Magazine*, June-July 1844.

11 J. Dickinson, *Statistics of Collieries of Lancashire, Cheshire and North Wales*, p. 72, published in *Memoirs* of the Manchester Literary and Philosophical Society (1855).

12 *The First Report of the Commissioner of Mines*, 1842, contained testimonies like that of Betty Harris, aged 37, who worked at Little Bolton coalpit: 'I have a belt round my waist and a chain passing between my legs, and I go on my hands and feet. The road is very steep ... The pit is very wet where I work, and the water comes over our clog-tops always, and I have seen it up to my thighs; it rains in at the roof terribly; my clothes are wet through almost all day long ... My cousin looks after my children in the day-time. I am very tired when I get home at night; I fall asleep sometimes before I get washed ... I have drawn till I have had the skin off me; the belt and chain is worse when we are in the family way. My feller (husband) has beaten me many times for not being ready. I were not used to it at first, and he had little patience: I have known many a man beat his drawer.'

13 Ibid., pp. 15-6. In the small Lancashire 'mountain mines', which had only a thin strata, the report stated that some children 'are so young they go in their bed-gowns'. At Oldham, many started work at the age of four while in Rochdale the more customary age was five or six.

14 Eric Hobsbawn, '*The British Standard of Living, 1790-1850*', an essay in his book, *Labouring Men*, p. 74. W. Cooke-Taylor in his book, *Notes on a Tour in the Manufacturing Districts of Lancashire*, also cites many instances of severe distress in 1841-2. In Stockport, where unemployment reached 50 per cent, the average weekly wage of those employed was 7s. 6¼d. and the average of those partly employed was 4s. 7¼d. (pp. 216-7).

15 William Beesley, who later became W. P. Roberts' clerk, was arrested with Feargus O'Connor and 56 Chartists in the prosecutions that followed the General Strike of 1842. He made the remark quoted in his testimony at the Lancaster Assizes.

16 *Wigan Gazette*, 22 January, 1841; *Liverpool Mercury*, 12 February, 1841.

17 *The Union*, September 1842.

18 The Manchester correspondent of the *North of England Magazine*, 26 September, 1842, admitted that Bolton spinners had won an increase while most workers had resumed work at their previous rates of pay and in a few instances with a trifling reduction.

19 *Wakefield Journal*, 11 November, 1842; Frank Machin, *The Yorkshire Miners*, p. 44. S. Webb in his book, *History of the Durham Miners*, p. 29, incorrectly states that the union was formed in 1841—not 1842—and suggests

that Martin Jude played a prominent part at the inaugural conference. I seems highly improbable that Jude even attended.

20 *Manchester Guardian*, 2 August, 1843.

21 I am indebted to Mr John Smethurst for drawing this point to m attention. See J. Smethurst, *Lancashire and the Miners' Association of Grea Britain and Ireland*, p. 6.

22 *Northern Star*, 5 August, 1843.

23 William Grocott, a Manchester weaver, was secretary of the South Lancashire Chartist District. For a time, he held that post whilst also bein secretary of the Lancashire Miners. (See J. Smethurst, op. cit., p. 10).

William Dixon, originally a pitman, later a weaver and owner of a Man chester temperance hotel, became a full-time official of the Miners' Associa tion. He fought the constituency of Wigan as a Chartist candidate in 1841 an 1847.

Henry Dennett was a veteran of the Lancashire colliers' union of 1831 He mentioned his experience in that earlier organization at the Mancheste conference, January 1844. John Berry had worked in ten coalfields an belonged to four unions prior to 1842. These included the Lancashire union of 1831 and 1834 as well as Hepburn's union of 1832. The *Miners' Advocate* of March 1847, which gave this information in a biographical sketch of Berry said that 'he is one of the main props of the Miners' Association'.

24 Ibid., 15 July, 1843.

25 Ibid., 19 August, 1843.

26 Ibid., 12 August, 1843. The altercation between Hulton and Swallow that appears on the following page is also taken from the same source.

27 Home Office 45. O.S. 350 1843.

28 Books on the subject include: Donald Read's *Peterloo*, Joyce Marlow's *The Peterloo Massacre*, and Robert Walmsley's *Peterloo: The Case Reopened*

29 Mr Robert Walmsley questions my interpretation. He suggests tha the person involved in the dispute with Swallow was not William Hulton bu his son, William Ford Hulton. (See his letters to *The Guardian* and my replies, 23 July, 1 August, 11 August and 18 August, 1969.) I argue that i is more likely to be William Hulton senior because: first, it would make nonsense of Swallow's remark about Hulton knowing more about Peterloo The son, W. F. Hulton, was a small boy at the time and almost certainly did not attend the fateful demonstration. And, second, it would be difficult to understand why, if it were the son, he should be sensitive about the subject ('Don you noa the consequence oa takin' abeawt Peterloo to th' Squoire?') nor why his servants should have taken such extreme action.

30 *Miners' Advocate*, March 1844. Lancashire circulation was handled by James Leach, the Chartist leader, from his premises in Broad Oak Street, Manchester.

31 W. P. Roberts at times tended to regard the strike weapon as a means, not merely of securing better pay, but of liberation. In an editorial for the *Miners' Journal*, October 1843, he tells colliers that a general strike will lead to 'your emancipation ... We deprecate partial strikes because they weaken us and thus postpone the hour of our social redemption.'

32 *Miners' Advocate*, 13 January, 1844.

33 Membership of the union at St Helens dropped from a thousand in January 1844 to 219 in November 1844. (Membership returns given to the

Manchester conference, January 1844, and to the Wakefield conference, November 1844.)

34 *Manchester Guardian*, 10 January, 1844, and 27 February, 1844.

35 *The Times*, 1 March, 1844.

36 *Miners' Advocate*, 6 April, 1844. Professor Barker in his book, *Pilkington Brothers and the Glass Industry*, makes it clear that the employers capitulated because of the shortage of coal. Supplies were exhausted in two or three weeks and, once the strike had ended, old workings were re-opened in an attempt to meet unfilled orders. It was because of the disruption caused to their glass works on this occasion that Pilkington Brothers decided to become big coal-owners and thus no longer be beholden to outside suppliers. (pp. 128-9.)

37 *Manchester Guardian*, 18 December, 1844.

38 William Cubbon, in his *Bibliography of Works relating to the Isle of Man*, says it must have ceased publication once the special postal privileges were withdrawn. It was published in Douglas from 1846 to 1849.

39 The entire national leadership was composed of Lancashire men by 1847. They exercised a decisive influence on decision-making well before that date.

40 *Northern Star*, 8, 15, 22, 29 June and 6 July; *Miners' Advocate*, 27 July, 1844. Leaflets calling for colliery collections to be made in Lancashire are included in the Pitmen's Strike Collection in Wigan Public Library.

41 *Bolton Chronicle*, 31 January, 1846. W. P. Roberts made this point when addressing a meeting at the beginning of the Bolton strike. Another speaker said that 'the children in the streets had now got clogs to their feet and more buttercakes than they did before'.

42 *Miners' Advocate*, 3 July, 1845, reported a typical demonstration, saying that 'the miners of Wigan are in a very prosperous state at present. One of the Masters lately gave a large piece of beef, bread and ale, and another two good barrels of ale, while a band of music played on the pit heap and flags fluttered in the breeze.'

43 Ibid., February 1847. A hundred men attended a union dinner at Bolton, where the employer, William Hulton, provided the money to buy all the refreshments. W. Daniels, editor of the *Miners' Advocate*, proposed a toast to 'William Hulton, Esq., and Son, God Bless them!'

44 *Bolton Free Press*, 4 April, 1846. A Prosecutor told the Bolton Quarter Sessions, 'In other cases fines have been inflicted, but I do not consider that a punishment since it was paid by the union.'

45 A. H. Dodd in his book, *The Industrial Revolution in North Wales*, pp. 343-4, says 'It seems, then, that from about 1830 to 1850 wages in North Wales' coal mines were at best not much more than three-quarters, and at worst barely half, those paid in England and Scotland.'

46 For vituperation against strikebreakers see, for example, the poem entitled 'The Nobstick (alias B . . .)', published in the *Miners' Advocate*, December 1847.

47 A meeting at the Ship Yard, Millgate, Wigan, to consider the respective merits of the National Association and the Miners' Association needed two chairmen and a neutral umpire to preserve order. *Northern Star*, 29 January, 1848.

48 *Northern Star*, 26 September, 1846. The relations between the Miners'

Association and the National Association are more fully dealt with in R. Challinor and B. Ripley, op. cit., pp. 196-200, pp. 230-4.

49 *Northern Star*, 2 May, 1846.

50 *Miners' Advocate*, February 1848. In his review of the book, *The Miners' Association: A Trade Union in the Age of the Chartists*, Dr J. E. Williams argues that R. Challinor and B. Ripley fail to establish the point that the union had close links with Chartism. See Bulletin no. 19 (Autumn 1969) of the Society for the Study of Labour History, pp. 39-40.

51 *Preston Chronicle*, 24 July, 1847.

52 *Northern Star*, 14 August, 1847; *Preston Pilot*, 31 July, 1847.

53 *Manchester Guardian*, 9 December, 1848.

54 Home Office papers (1849) O.S. 2,619.

NOTES ON CHAPTER THREE

1 *Red Republican*, 9 November, 1850.

2 *Friend of the People*, 25 February, 1851.

3 Select Committee on Accidents, 1954. Q.2466. His testimony before the Select Committee is the last reference to David Swallow I have encountered.

4 Rules of the Lancashire and Cheshire District of the Miners' Association, as revised and adopted by the county meeting at the Oddfellows Arms, Hurstbrook, Ashton-under-Lyne, on 29 March, 1852, is reprinted as Appendix Two of the Select Committee on Accidents, 1854. The Pitmen's Strike Collection at Wigan Public Library, the mining collection at Newcastle-on-Tyne Public Library, and Public Records Office H.O. 45 O.S. 434 all contain drafts of the original 1842 constitution.

5 *Wigan Observer*, 5 November, 1853.

6 Ibid. The conduct of the Wigan Police Force is dealt with more fully in the chapter entitled 'The Lancashire Miner and his Social Conditions'.

7 *Manchester Guardian*, 2 November, 1853.

8 W. W. Rostow, *British Economy of the Nineteenth Century*, p. 33.

9 C. Hardwick, *History of the Borough of Preston and its Environs*, pp. 418-9; A. Hewetson, *History of Preston*, p. 178; H. Ashworth, *The Preston Strike*. An Enquiry into its Causes and Consequences, *passim*.

10 *The People's Paper*, 12 November, 1853. Also, John Saville, Ernest Jones: *Chartist*, Lawrence & Wishart, London (1952) p. 203.

11 *Manchester Guardian*, 11 March, 1854.

12 *Northern Star*, 29 April, 1843.

13 A. J. Taylor, The Wigan Coalfield in 1851, *Transactions of the Historic Society of Lancashire and Cheshire*, vol. 106 (1954), p. 118.

14 J. Lunn, *Leigh: The Historical Past of a Lancashire Borough*, Leigh Corporation (1960) p. 147.

15 J. Dickinson, Statistics of the Collieries of Lancashire, Cheshire and North Wales (1854), published in *Memoirs of the Manchester Literary and Philosophical Society*, (1855), p. 71.

16 Report of the Royal Commission on Coal, 1871.

17 W. Fairley, *Calculations on the Produce of Coal Seams*, p. 5. Also, M. Dunn, in *Winning and Working of Collieries*, p. 36, mentions the premium placed on professional skill by this method of calculation.

18 The ill-fated attempts to run co-operative collieries in Lancashire are dealt with in more detail in chapter eight, 'Seven Lean Years'.

19 D. Anderson, Blundell's Collieries: Wages, Disputes and Conditions of Work, *Transactions of the Historic Society of Lancashire and Cheshire*, vol. 117 (1965), p. 138.

20 H. Macworth, On the Diseases of Miners, *The Journal of the Society of Arts*, 4 April, 1855.

21 J. Dickinson, op. cit., *Statistics of Colliers in Lancashire, etc.*, p. 75. He made the point that the death rate was much higher in Lancashire than in Belgium, where analogous conditions prevailed.

22 P. E. H. Hair, Mortality from Violence in British Coal Mines, 1800-50, *Economic History Review*, vol. xxi, December 1968, p. 546. He says on page 558 that deaths caused by explosions were twice the national average in Lancashire.

23 The Mines Inspector's Report on Lancashire, 1855, p. 19.

24 Select Committee on Coal Mines, 1852. Q.1060. Also the Mines Inspector's Report on Lancashire, 1852, pp. 44-6.

25 Quoted by M. Dunn, *A History of the Steam-Jet*, p. 5.

26 J. Dickinson, op. cit., *Statistics of Collieries of Lancashire, etc.*, partly attributed Lancashire's high mortality figures to the extensive use of machinery.

27 R. N. Boyd, *Coal Pits and Pitmen*, Whittaker & Co. London (1892) p. 212.

28 C. M. Percy, *Mining in the Victorian Era*, Wall, Wigan (1897) p. 52.

29 J. Dickinson, op. cit., *Statistics of the Collieries of Lancashire, etc.*, pp. 77-8.

30 *Daily News*, 24 April, 1868.

31 M. Kay, in her book, *The History of Rivington and Blackrod Grammar School*, makes no reference to this disparaging Report.

32 Report of the Commission appointed to inquire into the State of Popular Education in England, 1861, vol. i, p. 188.

33 R. Page Arnot, *The Miners: Years of Struggle*, Allen & Unwin, London (1953) vol. ii, p. 42.

34 Transactions of the National Society for the Promotion of Social Science, 1865, pp. 366-7.

35 *The British Miner*, 3 January, 1863.

36 M. Tylecote, *The Mechanics' Institute of Lancashire and Yorkshire before 1851*, p. 108.

37 *Manchester Guardian*, 1 September, 1849.

38 M. Tylecote, op. cit., p. 66.

39 This is more fully dealt with in the chapter, 'The Lancashire Miner and His Social Life'.

40 J. Dickinson, op. cit., *Statistics of the Collieries of Lancashire, etc.*, p. 72.

41 C. Colwell, *Fiery Facts*, p. 3.

42 J. Dickinson, op. cit., *Statistics of the Collieries of Lancashire, etc.*, p. 72.

43 Proceedings of the Miners' National Conference, Leeds, November 1873.

44 Alexander MacDonald, born 21 June, 1821, eldest of seven boys, entered the pits at the age of eight. He had to work for between fourteen and sixteen hours a day. Nevertheless, he went to evening school and, in 1846, attended Glasgow University. MacDonald took part in the colliers' strike of 1842 and, at the end of it, 'was selected as a victim for the tyranny of the employers'. For a while, he became a schoolteacher. In 1855, he was appointed general secretary of the Scottish Miners' Union. (Sources: R. Page Arnot, *The Scottish Miners*, and A. W. Humphrey, *A History of Labour Representation.*)

45 *Bolton Chronicle*, 26 May, 1860, reported another abortive attempt to form a union. The Report on Trade Societies Rules proposed for the National Association for the Promotion of Social Science (1860), by G. Shaw Lefevre, refers on page 138 to another union—the United Trades Association of Miners, Salt Makers and Boatmen of Lancashire and Cheshire. Its membership, he claims, consisted of 900 to 1,000 miners and 200-300 saltmakers. (p. 304.) I have been unable to discover any further references to this organization.

46 *Wigan Observer*, 22 October, 1864.

47 *Ibid.*, 8 July, 1865.

48 *Ibid.*, 10 October, 1862.

49 *Ibid.*

50 *The British Miner*, 10 January, 1863, and *Wigan Observer*, 6 February, 1863.

51 S. & B. Webb, *History of Trade Unionism, 1666-1920*, p. 262, and H. Pelling, *A History of British Trade Unionism*, pp. 58-9.

52 *Wigan Observer*, 25 November, 1865.

NOTES ON CHAPTER FOUR

1 *Wigan Observer*, 3 July, 1863.

2 *The Beehive*, October 1866.

3 The Lancashire miners appear to have often held their meetings at the Oddfellows Arms, Hurstbrook, Ashton-under-Lyne. Its premises were used by them when they passed the county union's revised constitution on 29 March, 1852. (See Chapter Three, Note 4.) Allen Tetlow was the landlord of the Oddfellows Arms and he must have been experienced in providing suitable conference facilities.

4 *Wigan Observer*, 14 May, 1864. See table 7: Date of Formation and Members of Lancashire District Unions.

5 Ashton-under-Lyne was later to show its independence in the 1870s and 1880s, by remaining affiliated to the Miners' National Union when all other Lancashire district unions disaffiliated.

6 *The Miners' and Workmen's Advocate*, 9 January, 1864.

7 *Wigan Observer*, 14 May, 1864.

8 *The Miners' and Workmen's Advocate*, 27 February, 1864.

9 J. Evans' testimony to the R.C. on Trade Unions, 28 April, 1868. Q.15225.

10 *The Beehive*, 6 August, 1864.

11 J. B. Smethurst, *Strikes and Strikebreakers in the Worsley Coalfield*, p. 11.

12 *Wigan Observer*, 13 January, 1865.

13 By the beginning of 1866, Farnworth and Kearsley district union's bank balance had risen to £2,000. See Halliday's statement in *The Commonwealth*, 24 February, 1866.

14 *The Miner*: A Journal for Underground Workers, edited by J. K. Hardie, May 1883.

15 *Wigan Observer*, 26 August, 1865.

16 *The Miners' and Workmen's Advocate*, 29 October, 1864.

17 *Wigan Observer*, 23 January, 1869.

18 *The Miners' and Workmen's Advocate*, 30 April, 1864.

19 W. Pickard's testimony to the R.C. on the Scarcity and Dearness of Coal, 1873. Q.4210.

20 *The Beehive*, 2 May, 1873.

21 Ibid.

22 *The Miners' and Workmen's Advocate*, 26 March, 1864.

23 *Bristol Times*, 15 October, 1873.

24 *Wigan Observer*, 9 October, 1869.

25 Ibid., 21 November, 1863. Also, see the discussion of educational provisions in the Lancashire coalfield in Chapters three and fifteen.

26 *Bolton Evening News*, 26 June, 1867.

27 *Wigan Observer*, 9 October, 1869.

28 Ibid., 21 November, 1863.

29 *The British Miner*, 21 March, 1863.

30 *Wigan Observer*, 13 January, 1865.

31 Andrew Wynter, *Peeps into the Human Hive*, 'My Lords of Coal', pp. 62-3.

32 *Wigan Observer*, 7 July, 1871.

33 South Lancashire and Cheshire Coal Association, annual general meeting, Manchester, 1877. It may be that a further reason for the authorities wanting to restrict the miners' access to gunpowder was that it was sometimes used in industrial disputes.

34 J. M. Forster, 'The Pit Brow Lassie', serialized in the *Wigan Comet*, 12 February, 1889.

35 John Plummer, *A Real Social Evil*, p. 3.

36 *Daily News*, 1 May, 1868.

37 *Wigan Comet*, 26 January, 1889.

38 W. Pickard's testimony to R.C. on the Scarcity and Dearness of Coal, 1873. The heroine of J. M. Forster's 'The Pit Brow Lassie' marries a collier from Up Holland at the end of the novel.

39 *An order for the Pit*, a description of Fletcher, Burrows and Co., Atherton, p. 23.

40 J. M. Forster, op. cit., *Wigan Comet*, 9 February, 1889.

41 Quoted John Plummer, op. cit., p. 3.

42 *Bolton Chronicle*, 24 November, 1866. Also T. Halliday's testimony to R.C. on Trade Unions, 1868. Q.16688.

43 T. Halliday testimony to R.C. on Trade Unions, 1868. Q.16654.

44 Letter to Messrs Andrew Knowles and Son, from the Pendlebury Committee of the Union, dated 12 December, 1866.

45 *Bolton Chronicle*, 15 December, 1867. Also, R.C. on Trade Unions, 1868, evidence of John Knowles. Q.16238-44.

46 Ibid.

47 *Bolton Chronicle*, 23 February, 1867. Also, R.C. on Trade Unions, 1868, evidence of John Knowles. Q.16238-44.

48 *Swinton and Pendlebury Journal*, 21 March, 1962. Article by C. J. Collier. Also, J. B. Smethurst, *Strikes and Strikebreakers*, op. cit., p. 15.

49 *Bolton Chronicle*, 2 March, 1867.

50 Ibid., 17 and 23 March, 1867. Also, John Knowles' testimony to R.C. on Trade Unions, 1868, Q.16285-92, and J. Halliday testimony Q.16691-3.

51 J. Halliday's testimony to R.C. on Trade Unions, 1868. Q.16691-2. Part of his answer is quoted in the next chapter.

52 Joseph Bent, *Criminal Life*: Reminisences of Forty-two years as a Police Officer, p. 57. It is interesting to note that a man named Bent (informer B), of Manchester, was employed as a police spy on working class organization in the first part of the nineteenth century. I have been unable to establish that the two Bents are in any way related.

53 *Bolton Chronicle*, 23 February, 1867.

54 William Holford's testimony to R.C. on Trade Unions, 1868. Q.16411-16444.

55 *Swinton and Pendlebury Journal*, 21 March, 1962.

56 K. Knowles' testimony to R.C. on Trade Unions, 1868. Q.16301.

57 *Bolton Chronicle*, 29 June, 1867.

58 Alexander MacDonald's testimony to R.C. on Trade Unions, 1868. Q.15003.

59 Ibid. Q.15021.

60 *Wigan Observer*, 13 March, 1868.

61 Ibid. Also the Count de Paris in his book, *The Trade Unions of England*, stated that the union leaders 'declared that they used their utmost efforts, so far as even to endanger their influence, in an endeavour to enforce moderation and to calm the passions'.

62 J. Evans' testimony to R.C. on Trade Unions, 1868. Q.15152.

63 William Rowe's testimony to R.C. on Trade Unions, 1868. Q.14918.

64 *Wigan Observer*, 27 March, 1868.

65 Alfred Hewlett's testimony to R.C. on Trade Unions, 1868. Q.15044.

66 Ibid. Q.15104.

67 *Wigan Observer*, 3 April, 1868, and Alfred Hewlett's testimony, op. cit.

68 Alfred Hewlett's testimony, op. cit. Q.15106.

69 *Wigan Observer*, 24 April, 1868.

70 *The Spectator*, 10 April, 1868.

71 *Wigan Observer*, 7 May, 1868.

72 Leaflet published 28 March, 1880. Contained in *Wigan Broadsides*, vol. i, (Wigan Public Library).

73 *Manchester Courier*, 5 May, 1868.

74 *Wigan Observer*, 15 August, 1868. The miners appear to have been preoccupied with their own worries in 1868 and did not play a prominent part in the inaugural conference of the T.U.C. at Manchester. See A. E. Musson, *The Congress of 1868*.

NOTES ON CHAPTER FIVE

1 *Wigan Observer*, 1 May, 1869.
2 *Bolton Chronicle*, 26 June, 1869.
3 *Wigan Observer*, 10 July, 1869.
4 Ibid., 24 April, 1869.
5 S. & B. Webb, *History of Trade Unionism*, pp. 233-98; H. Pelling, *A History of British Trade Unionism*, pp. 59-64; H. A. Turner, *Trade Union Growth, Structure and Policy*, pp. 215-220.
6 *Wigan Observer*, 3 June, 1865.
7 *The Beehive*, 3 June, 1865.
8 *Bolton Evening News*, 27 June, 1867.
9 T. Halliday's testimony to the R.C. on Trade Unions, 1867. Q.16693.
10 *Wigan Observer*, 30 July, 1869.
11 G. D. H. Cole, Some Notes on British Trade Unionism in Essays in Economic History (ed. E. M. Carus-Wilson) vol. iii, p. 209.
12 *Wigan Observer*, 28 August, 1869.
13 Ibid.
14 *Bolton Chronicle*, 12 February, 1870.
15 Ibid., 19 February, 1870.
16 *Wigan Observer*, 9 October, 1869
17 *Wrexham Advertiser*, 2 October, 1869.
18 Ibid., 11 April, 1863.
19 *Wigan Observer*, 28 August, 1869.
20 Ibid., 6 January, 1870.
21 Ibid.
22 Ibid.
23 Quoted G. D. H. Cole, op. cit., p. 209.
24 *Wrexham Advertiser*, 22 April, 1870.
25 H. A. Turner, *Trade Union Growth, Structure and Policy*, p. 215.
26 *Wigan Observer*, 13 January, 1871.
27 Ibid., 5 May, 1871.
28 Select Committee on Mines, 1866. Q.4373.
29 *Wigan Observer*, 8 November, 1872.
30 G. D. H. Cole, op. cit., p. 210.
31 *Wigan Observer*, 19 September, 1873.
32 Ibid., 10 October, 1873.
33 Ibid., 6 January, 1870, and 9 September, 1871.

NOTES ON CHAPTER SIX

1 *Wigan Observer*, 22 April, 1870.
2 The South Wales miners' pamphlet, '*The Double and Single Shift Compared*', contains the petition to Parliament, presented on 16 May, 1868, asking for workmen to be appointed as sub-inspectors.
3 *Wigan Observer*, 22 April, 1870. Halliday attacked MacDonald, Pickard and Normansell in his presidential address to the Wrexham conference.

4 A Dalziel, *The Colliers' Strike in South Wales.* Its Causes, Progress and Settlement, *Western Mail,* Cardiff (1872), p. 28.

5 Ibid., p. 30.

6 *The Beehive,* 11 February, 1871; E. W. Evans, *The Miners of South Wales,* p. 101.

7 A. Dalziel, *The Colliers' Strike in South Wales,* p. 30.

8 *Wigan Observer,* 14 July, 1871.

9 Ibid., 5 May, 1871.

10 Ibid., 28 April, 1871.

11 Ibid., 8 April, 1871.

12 Ibid., 5 May, 1871.

13 Ibid., 22 April, 1870.

14 A. Dalziel, *The Colliers' Strike in South Wales,* p. 12.

15 *Western Mail,* 18 July, 1871. The A.A.M.'s national membership figures were given to the Manchester conference, 1 May, 1871.

16 A. Dalziel, *The Colliers' Strike in South Wales,* p. 37.

17 Ibid., p. 30.

18 Ibid., p. 35.

19 J. H. Morris, and L. J. Williams, *The South Wales Coal Industry,* University of Wales Press, Cardiff (1958) p. 278.

20 A. Dalziel, *The Colliers' Strike in South Wales,* p. 44.

21 Ibid., p. 107.

22 Ibid., p. 108.

23 A. Dalziel, *The Colliers' Strike in South Wales,* p. 127.

24 *Western Mail,* 20 August, 1871.

25 E. W. Evans, *The Miners of South Wales,* p. 105.

26 V. L. Allen, The Origins of Industrial Conciliation and Arbitration, *International Review of Social History,* Vol. IX (1964), Part 2, p. 249.

27 *The Times,* 18 April, 1868.

28 Ibid., 5 May, 1868.

29 *Colliery Guardian,* 24 July, 1868.

30 Ibid.

31 Ibid.

32 E. W. Evans, *The Miners of South Wales,* p. 107.

33 Ibid.

34 R. Page Arnot, *The Miners,* Vol. 1, p. 69.

35 *Wigan Observer,* 28 April, 1871.

36 *The Beehive,* 19 April, 1873.

37 E. W. Evans, *The Miners of South Wales,* p. 108.

38 George Elliot, an owner with mines in Northumberland, North and South Staffordshire, North and South Wales, testimony to the R.C. on the Scarcity and Dearness of Coal. Q.7549.

39 *Manchester Guardian,* 18 April, 1873.

40 Thomas Cooper, the old Chartist leader, visited the Lancashire Coalfield about this time and wrote the following description in his autobiography: 'My sorrowful impressions were confirmed. In our old Chartist time, it is true, Lancashire working men were in rags by the thousands; and many of them often lacked food. But their intelligence was demonstrated wherever you went. You would see them in groups discussing the great doctrine of political justice ... *Now* you will see no such groups in Lancashire. But you

will hear well-dressed working men talking, as they walk with their hands in their pockets, of "Co-ops" and their shares in them, or in building societies. And you will see others, like idiots, leading small greyhound dogs.' *The Life of Thomas Cooper*, written by Himself, (1872), p. 393.

NOTES ON CHAPTER SEVEN

1 *Wigan Observer*, 7 March, 1873.
2 K. Marx, *Capital*, Vol. III, p. 72.
3 Ibid., p. 73.
4 A. Hewlett's testimony to S.C. on the Scarcity and Dearness of Coal, 1873. Q.2020.
5 J. Dickinson's testimony to S.C. on the Scarcity and Dearness of Coal, 1873. Q.257.
6 J. Dickinson, ibid. Q.302.
7 William Frith's address to the British Association at Bradford, 21 September, 1873.
8 James S. Jeans in the *Practical Magazine*, April 1874. The advantages accruing from the introduction of coal-cutting machinery may well have increased not simply as the models became more reliable but also because extensive use led to reductions in cost. The cost of a coal-cutting machine in 1853-4 was about £500; by 1862-3 the figure was only £50 plus the cost of pipes. (Proceedings of the North of England Mining Engineers, 1853-4, vol. ii, p. 72, and Proceedings of the North of England Mining Engineers, 1862-3, vol. xii, a report of the extensive use of machinery at Ince Hall Collieries.)
9 *Colliery Guardian*, 4 August, 1875.
10 H. S. Jevons, *The British Coal Trade*, p. 73.
10a J. Dickinson's testimony, ibid.
11 R. N. Boyd, *Coal Pits and Pitmen*, p. 240.
12 *Wigan Observer*, 25 April, 1874.
13 Ibid., 30 May, 1872.
14 R. Hunt, 1854 *Mining Records* and *Wigan Observer*, 25 April, 1874.
15 *Wigan Observer*, 26 April, 1874.
16 J. Dickinson's testimony to S.C. on Scarcity and Dearness of Coal, 1873. Q.144.
17 W. Bennett, *History of Burnley*, Burnley Corporation (1951) vol. iv, p. 109.
18 A. Hewlett's testimony to the Select Committee on the Scarcity and Dearness of Coal, 1873. Q.2021.
19 T. Halliday's testimony, ibid. Q.5216.
20 T. Halliday's testimony, ibid. Q.5276.
20a G. Elliot's testimony, ibid. Q.7542.
21 *Wigan Observer*, 15 February, 1873.
22 *Colliery Guardian*, 4 October, 1872, quoting Amalgamated Association's Walsall conference report.
23 *Wigan Observer*, 5 April, 1872.
24 Ibid.
25 Ibid., 24 January, 1874.

26 Full Report of the Great Trial for Conspiracy, 18-20 March, 1874, published by A.A.M., Burnley (W. Smith, 20 Hargreaves Street).

27 *Labour Press and Workmen's Examiner*, 4 April, 1874.

28 *Wigan Observer*, 9 May, 1874.

29 Ibid., 2 October, 1874.

30 Ibid.

31 *Manchester Examiner*, 9 October, 1874.

32 *Wigan Observer*, 10 October, 1874.

33 Editorial comment in *Wigan Observer*, 17 October, 1874.

34 *Wigan Observer*, 31 October, 1874.

35 James Wilson's letter, *Wigan Observer*, 21 November, 1874.

36 Correspondence of Lord Aberdare and T. Halliday quoted *Wigan Observer*, 19 February, 1875.

37 *Wigan Observer*, 26 February, 1875.

38 N. Edwards, *South Wales Miners*, Lawrence & Wishart, London (1938) pp. 68-9.

39 In his book, '*Mabon. A study in Trade Union Leadership*', Mr W. E. Evans remarks that 'Mabon's rise from collier to miner's agent was impressive if only for its amazing rapidity', p. 10. Mr Evans continues his account by contrasting Mabon's outlook with that of Halliday: 'Halliday was less cautious than Mabon, and not so unwilling to use the strike weapon. He was an able, enthusiastic leader who believed there could be no compromise on a matter of principle. Mabon, on the other hand, was appalled by the hardships suffered by strikers and their families during the stoppages of 1875, and as the months passed he found himself in opposition to Halliday.' (p. 11.)

40 *The Beehive*, 17 July, 1875.

41 *Wigan Observer*, 27 August, 1875.

NOTES ON CHAPTER EIGHT

1 *Wigan Observer*, 9 September, 1871.

2 W. E. Gladstone letter to Enoch Robinson, 20 September, 1871, reprinted in the *Wigan Observer*, 29 September, 1871.

3 *Colliery Guardian*, 22 September, 1871.

4 Professor Beesly's letter to T. Halliday, reprinted in the *Wigan Observer*, 13 October, 1871. Beesly's condemnation of MacDonald's association with Lord Elcho should be considered in the context of Lordship's Parliamentary speeches. When the Reform Bill came before Parliament, he 'called upon the House to "resist the democratic tide which was sitting in" while there was still time'. (Quoted F. E. Gillespie, *Labour and Politics in England, 1850-1867*, p. 175.) He also called upon the government to take strong action against combinations of workmen because of the tyranny of trade unionism, "which of all tyrannies was the most grinding and intolerable'. (Ibid., p. 177.)

Nevertheless, Alexander MacDonald frequently accepted invitations to champagne breakfasts and social functions with His Lordship. (Earl of Wemyss and March, Memories 1818-1912, cited by Royden Harrison, *Before the Socialists*, Routledge & Kegan Paul, London (1965) p. 119.

5 *The Beehive*, 24 August, 1872.

6 *Marx-Engels on Britain*, Foreign Languages Publishing House, Moscow (1955) p. 419.

7 *Kilmarnock Standard*, 7 September, 1868. Beesly had previously assisted MacDonald in other moves. In his essay, Professor Beesly and the Working-class Movement, Dr Royden Harrison writes: 'While Alexander MacDonald was struggling to establish his new National Association of Miners, Beesly worked on a sub-committee in London which helped MacDonald to secure a parliamentary inquiry into conditions in the mines and which assisted the miners' leader in his battle with the disruption tendencies inside the union which were being exploited by a journal called the *British Miner*.' ('Essays in Labour History' (ed. Briggs and Saville), p. 214.)

8 *The Beehive*, 3 August, 1872.

9 R. Page Arnot, *A History of the Scottish Miners*, p. 53.

10 *Wigan Observer*, 27 June, 1873. Rank-and-File restiveness was also manifest by public meetings held in Lancashire to answer the question: 'Did the Miners' Parliamentary deputation betray the interest of the constituents while in London and forfeit the confidence of the mining population for the future?' Two handbills advertising meetings in the Oldham district on this topic are contained in the Correspondence of E. L. Stanley. John Rylands library. Eng MS 1095.

11 *The Beehive*, 20 July, 1872. Beesly's view was sometimes echoed in print by working miners. Samuel Broome, in a letter to the *Oldham Examiner and Times*, 21 November, 1872, wrote: 'It is, I think, needless for agents to argue that they did the best they could under the circumstances. The "best" they could have done was to have kept aloof from the coalowners, and gone in for what they had previously so strenuously advocated, by supporting Mr. Bruce.'

12 A. MacDonald, *The Mines Regulation Act*, published Glasgow, 1872. John C. Fowler's book, *Collieries and Colliers*. A Handbook of the Law and Leading Cases Relating thereto, gives a more comprehensive picture of the legal position after the passing of the 1872 Act.

13 H. A. Bruce, *Letters of Lord Aberdare* (printed for private circulation), cited Dr R. Harrison, *Before the Socialists*, p. 297.

14 *The Beehive*, 3 August, 1872.

15 Mundella-Leader Correspondence, cited J. E. Williams, *The Derbyshire Miners*, p. 140. Historians disagree over their verdict on Alexander MacDonald. The majority view, exemplified by the Webbs, G. D. H. Cole and Page Arnot, is that his influence was, generally speaking, highly beneficial to the miners' cause. The main dissidents appear to be Professor Youngson-Brown, who confines his study of MacDonald to the industrial policy he pursued in the Scottish coalfields, and Professor Royden Harrison, who has studied his political activity in the 1860s and 1870s.

Dr J. E. Williams takes an intermediary position, arguing that 'Mr. Page Arnot rightly disputes Professor Youngson-Brown's assumption that MacDonald's "class harmony" standpoint of the 'seventies was also fully typical of MacDonald in his earlier years'. (*Bulletin 4 of the Society for the Study of Labour History. Spring 1962.*)

As this chapter indicates, I tend to agree with Professor Youngson-Brown and Professor Harrison.

16 W. B. Gwyn, *Democracy and the Cost of Politics*, p. 150, and R. Harrison, *Before the Socialists*, p. 175.

17 *Wigan Observer*, 22 November, 1873. Once MacDonald was returned for Stafford, his financial arrangements give rise to some perplexity. In his book '*The New Parliament*, 1880' W. Saunders says MacDonald's address when Parliament was sitting as Westminster Palace Hotel, S.W. Harry McShane, author of '*The History of Glasgow Trades Council*', in an interview with me said that MacDonald had a large residence on the outskirts of Glasgow.

Presumably his way of life, both in Parliament and at home in Scotland, required an income of considerable size. The question is: Where did it come from?

18 Professor B. C. Roberts, *The Trades Union Congress*, 1868-1921, says that Congress delegates were surprised when the winner's name was announced. It was John Wilson of Sheffield, 'who was notorious for his written and spoken utterances against them (i.e. trade unions)' (p. 80).

19 A. MacDonald's testimony to R.C. on Trade Unions, 29 April, 1868.

20 *Wigan Observer*, 3 September, 1875.

21 *Wigan Observer*, 5 April, 1872.

22 W. B. Gwyn, *Democracy and the Cost of Politics*, London University Press, London (1962) p. 151.

23 *Wigan Observer*, 22 November, 1873.

24 R. Page Arnot, *The Miners*, vol. 1, p. 291.

25 G. D. H. Cole, *British Working Class Politics, 1832-1914*, Routledge, London (1941) p. 56.

26 *Wigan Observer*, 11 October, 1872.

27 *Wigan Observer*, 10 October, 1873.

28 *Wigan Observer*, 18 October, 1872.

29 R. M. Latham's obituary in *The Beehive*, 2 May, 1873, and *Wigan Observer*, 9 May, 1873.

30 G. D. H. Cole, *British Working Class Politics, 1832-1914*, p. 50.

31 *Wigan Observer*, 24 January, 1874.

32 Howell Collection. Election Report of Sanders and Wade.

33 D. Brown, *The Politics of Wigan, 1874-1950*. First chapter 'A Tory Working Class'. (Liverpool University, M.A. thesis.)

34 H. J. Hanham, *Elections and Party Management*, Longmans, London (1959) pp. 105-6.

35 R. T. MacKenzie and Allan Silver, *Conservativism, Industrialism and the Working Class Tory in England.* (A paper presented to the Fifth World Congress of the International Sociological Association, Washington, 1962.)

36 *Wigan Observer*, 5 January, 1872.

37 *Wigan Observer*, 31 January, 1873.

38 *Wigan Examiner*, 24 January, 1874.

39 H. Pelling, *Social Geography of British Elections, 1885-1910*, Macmillan, London (1967) p. 267.

40 *Wigan Observer*, 31 January, 1874.

41 W. B. Gwyn, *Democracy and the Cost of Politics*, p. 150.

42 B. Disraeli's speech at Buckingham, 10 February, 1874.

43 *Marx-Engels on Britain*, pp. 469-470.

NOTES ON CHAPTER NINE

1 In his pamphlet, *Address to Miners of the United Kingdom* (1877), Alexander MacDonald explained that production of coal had risen from 125m. tons in 1874 to 131m. in 1875 and 134m. in 1876. But the increased output was accompanied by a drop in selling price of four shillings, which caused reduced incomes for all associated with the industry.

2 T. Ashton, *Three Big Strikes in the Coal Industry*, Vol. 1, p. 4. N. Stringfellow's speech, reported *Wigan Observer*, 29 January, 1881.

3 *Wigan Observer*, 8 July, 1876.

4 *Manchester Examiner*, 17 November, 1870.

5 *Wigan Observer*, 20 July, 1877. In other disputes, such as those of 1868 and 1881, the same pattern prevailed: non-union men were among the most intransigent.

6 *Wigan Observer*, 11 May, 1877.

7 *Wigan Observer*, 22 June and 29 June, 1877.

8 *Labour Press & Workmen's Examiner*, 20 November, 1875.

9 *Merthyr Tydvil Express*, 23 June, 1876; *Manchester Guardian*, 4 August, 1875.

10 *Wigan Observer*, 24 August, 1877.

11 *Wigan Observer*, 18 May, 1878.

12 *Wigan Observer*, 30 October, 1880.

13 R. Page Arnot, *The Miners*, Vol. 1, p. 82. In another of his books, Mr Arnot suggests that it was partly Halliday's influence that led William Brace to struggle against Mabon and the sliding scale in the latter part of the nineteenth century. (See R. Page Arnot, *South Wales Miners*, p. 30.)

14 *Wigan Observer*, 6 September, 1879.

15 T. Burt's letter in *Wigan Observer*, 28 December, 1877.

16 *Wigan Observer*, 1 June, 1877.

17 E. Hobsbawm, *Industry and Empire*, p. 101. I endeavour to explain why large Lancashire coalowners did not become 'new model' employers in chapter sixteen.

18 H. A. Turner, *Trade Union Growth, Structure and Policy*, p. 190.

19 *Wigan Observer*, 19 May, 1876.

20 B. Jones, *Co-operative Production*, p. 503.

21 Ibid.

22 Ibid., p. 505.

23 *Wigan Observer*, 14 September, 1877, and B. Jones, op. cit., p. 523.

24 *Co-operative News*, Vol. 5, pp. 71, 116.

25 B. Jones, op. cit., p. 539.

26 In his book, *A Short History of the British Working-Class Movement*, G. D. H. Cole said that trade unions started a co-operative mine at Oldham in the 1840s (p. 114). I have been unable to trace any reference to this mine in Lancashire records.

27 *Wigan Observer*, 5 April, 1873.

28 *Wigan Observer*, 2 March, 1877.

29 See *Report of First General Meeting of the Lancashire & Cheshire Miners' Permanent Relief Society*, held 10 March, 1873. Published by Wall Printer, Wigan.

30 G. L. Campbell, *Miners' Insurance Funds—their origin and extent*, p. 10.

31 Figures given in *Colliery Accidents in the Wigan District*, Report of the Relief Committee, September 1871.

32 *Wigan Observer*, 10 May, 1879.

33 *Wigan Observer*, 16 January, 1869.

34 *Colliery Guardian*, 30 January, 1869.

35 *Sheffield Telegraph*, 1 January, 1869.

36 *Wigan Observer*, 10 April, 1869.

37 *Wigan Observer*, 31 July, 1869.

38 *The British Miner*, 7 February, 1863.

39 *Wigan Observer*, 19 May, 1879. John Howcroft, Sen., in his Address to the Coal Miners of Ashton-under-Lyne and its vicinity, 1882, said he had worked in the pits for fifty-five years and been injured on twenty occasions.

40 *Wigan Observer*, 19 May, 1879.

41 Knowles' speech quoted *Wigan Observer*, 2 June, 1876.

NOTES ON CHAPTER TEN

1 T. Ashton, *Three Big Strikes in the Coal Industry*, vol. 1, p. 5 and R. Page Arnot, *The Miners*, vol. 1, p. 63.

2 J. Dickinson, *The Inspectors of Mines Report, 1881.*

3 *Manchester Guardian*, 5 March, 1881.

4 West Lancashire Coalowners' statement of 14 January, 1881, published *Wigan Observer*, 15 January, 1881.

5 *Leigh Chronicle*, 5 February, 1881.

6 Ibid., 29 January, 1881.

7 *Wigan Observer*, 8 January, 1881, and *Bolton Evening News*, 7 January, 1881.

8 *Wigan Observer*, 15 January, 1881.

9 Ibid.

10 Ibid., 8 January, 1881.

11 *Leigh Chronicle*, 5 February, 1881.

12 *Wigan Observer*, 16 April, 1881.

13 Resolution passed by a public meeting in Wigan Market Square, 4 February, 1881, reported the *Wigan Observer*, 5 February, 1881.

14 J. M. Forster's novels included *Children of Darkness* and *A Pit-Brow Lassie.*

15 *Wigan Observer*, 29 January, 1881.

16 Ibid., 5 February, 1881.

17 Ibid., 12 February, 1881.

18 *Manchester Guardian*, 28 January, 1881. Bernard Blakeley, in his booklet, *A History of St. Stephen's School, Kearsley*, quotes the following entry in the school log book for 27 January, 1881: 'On account of general distress owing to the colliery strike, the children were this day supplied with dinner in school. Also spoke to the vicar about school fees. Many arrears on account of distress.'

19 *Leigh Chronicle*, 11 January, 1881.
20 Ibid., 8 January, 1881.
21 *Wigan Observer*, 19 January, 1881.
22 *Leigh Chronicle*, 22 January, 1881.
23 *Wigan Observer*, 15 January, 1881.
24 Ibid., 29 January, 1881. At Ellesmere pit strikers were distributing cutlasses from a bag when Bent, a police officer, arrived. He was detested by miners because of his conduct in previous struggles. So the shout went up, 'Throw him down the pit.' Bent extricated himself from the dangerous situation by telling the other policemen present to take a good look at the colliers' faces; those who murdered him would be brought to trial.
25 *Leigh Chronicle*, 29 January, 1881, and *Bolton Evening News*, 25 January, 1881.
26 *Wigan Observer*, 21 May, 1881, and *Bolton Evening News*, 15 May, 1881. Reports of Ince Hall Coal and Cannel Company's application for compensation before the Liverpool Assizes, 14 May, 1881.
27 Letter from General Cambron to Lancashire Annual Sessions, quoted *Wigan Observer*, 1 April, 1881.
28 *Leigh Chronicle*, 5 February, 1881.
29 Ibid.
30 *Bolton Evening News*, 4 February, 1881.
31 *Leigh Chronicle*, 5 March, 1881.
32 *Hansard*, vol. CCLVII. Questions raised in the House of Commons, 27 and 28 January, 1, 8 and 15 February, 1881.
33 *Wigan Observer*, 26 February, 1881.
34 Ibid., 29 January, 1881.
35 T. Ashton, *Three Big Strikes in the Coal Industry*, vol. 1, p. 5, and W. Hallam, *Miners' Leaders*, p. 49.
36 *Wigan Observer*, 19 February, 1881.
37 W. Hallam, *Miners' Leader*, p. 50. Also R. Page Arnot, *The Miners*, vol. 1, p. 63.
38 *Wigan Observer*, 28 June, 1881.
39 Ibid., 20 April, 1878.
40 Ibid., J. Winstanley's letter, 2 July, 1881.

NOTES ON CHAPTER ELEVEN

1 *Wigan Observer*, 26 May, 1881. It is impossible to say whether this assertion is correct or not. No union balance sheets or records from this period survive. This itself can be taken as a probable indication of organizational weakness. The few, scanty references to union activity in local newspapers give, as suggested in the text, an impression of weakness and disunity.
2 Ibid., 2 July, 1881.
3 Ibid., 16 July, 1881.
4 Ibid., 13 March, 1880 and 23 July, 1881.
5 Ibid., 7 May, 1881.
6 T. Ashton, *Three Big Strikes in the Coal Industry*, vol. 1, p. 12.
7 Quoted R. Page Arnot, *The Miners*, vol. 1, pp. 69-70.

8 R. Page Arnot, ibid.

9 Lancashire Miners' Federation Committee meeting, 23 June, 1886; L.M.F. Conference, 31 August, 1886; T. Ashton, op. cit., vol. 1, p. 46. The fact that North Staffordshire was permitted to become so behind with its subscriptions would tend to show that the L.M.F. were reluctant to lose their links with the Staffordshire men and deprive themselves of the advice of Enoch Edwards.

10 T. Ashton, op. cit., vol. 1, p. 29. Trade depression made the time unpropitious for a national strike.

11 *Justice*, 5 and 12 July, 1884.

12 *Lancashire Miners' Federation Journal*, 24 November, 1883. Almost all issues of this journal, which would be so valuable to the writing of a history of the Lancashire miners, appear to have been lost.

13 T. Ashton, op. cit., vol. 1, p. 37.

14 Thomas Ashton wrote to the Mining Association in February and April 1882 and in April and June 1883.

15 *Wigan Observer*, 10 July, 1883.

16 L.M.F. half-yearly reports for 1886. Also, T. Ashton, op. cit., vol. 1, p. 33.

17 Quoted by Professor A. J. Youngson-Brown, Trade Union Policy in the Scots Coalfield, 1855-1885, *Economic History Review*, vol. VI, April 1954.

18 *Wigan Observer*, 9 June, 1886.

19 A. J. Taylor, Labour Productivity and Technological Innovation in the British Coal Industry, 1850-1914, *Economic History Review*, Second Series, vol. XLV, August 1961, p. 51.

20 Ibid., p. 57.

21 Ibid., p. 48.

22 *Nineteenth Century*, December 1884.

23 S. & B. Webb, *Industrial Democracy*, p. 557.

24 S. & B. Webb, ibid., p. 576.

25 H. S. Jevons, *The British Coal Industry*, p. 364.

26 R. Page Arnot, *The Miners*, vol. 1, p. 61.

27 J. W. F. Rowe, *Wages in the Coal Industry*, p. 62.

28 T. Ashton, op. cit., vol. 1, p. 65.

29 Pit-set conference, 2 February, 1888, also L.M.F. conference, 26 November, 1887.

30 John R. Rayner, *Coal and its Conflicts*, p. 142.

31 L.M.F. conference, 26 November, 1887.

32 Meeting between representatives of the employers and the L.M.F. 26 June, 1887.

33 T. Ashton, op. cit., vol. 1, p. 68.

34 Ibid., p. 81.

35 Ibid., p. 88.

NOTES ON CHAPTER TWELVE

1 E. Bainbridge, *The Coal Strike of 1893, The Contemporary Review, 1894*. (See Table 15.)
2 Thomas Ashton, *Three Big Strikes in the Coal Industry*, vol. 1, p. 139.
3 Thomas Ashton, ibid., p. 139-140.
4 H. A. Clegg, A. Fox and A. F. Thompson, *A History of British Trade Unions since 1889*, Oxford University Press (1964) p. 107.
5 R. Page Arnot, op. cit., vol. i, p. 205.
6 Thomas Ashton, op. cit., vol. i, p. 242.
7 *Leigh Chronicle*, 2 February, 1894.
8 L.M.F. Report, 8 April, 1890.
8a *Leigh Chronicle*, 9 November, 1894.
9 Thomas Ashton, op. cit., vol. ii, p. 78.
10 L.M.F. Conference, 21 January, 1888.
11 L.M.F. Conference, 19 August, 1893.
12 L.M.F. Conference, 2 June, 1894.
13 L.M.F. Conference, 12 January, 1895.
14 Thomas Ashton, op. cit., vol. ii, p. 59; L.M.F. Conference, 11 September, 1894.
15 Ibid.
16 L.M.F. Conference, 29 June, 1895, and 14 December, 1895.
17 Aspull district was re-admitted to the Federation by the conference of 21 May, 1889.
18 Thomas Ashton, op. cit., vol. ii, p. 61.
19 *The Comet*, 13 July, 1889.
20 Ibid., 2 November, 1889.
21 *The Pendlebury Pioneer*, May 1889.
22 *Labour Tribune*, 21 June, 1890.
23 L.M.F. Resolution, 8 April, 1890.
24 Thomas Ashton, op. cit., vol. ii, p. 123.
25 T. Ashton's letter to Henry Bramall, 19 November, 1891.
26 *Workmen's Times*, 6 February, 1892.
27 Bridgwater Collieries Ltd., General Accounts, 1844.
28 Royal Commission on Labour, C-6708-IV, 1892, Qs.5867, 5971.
29 *Justice*, 29 October, 1892.
30 J. Whitaker and T.T. Wilkinson, *Abstract of the Principal Mines of the Burnley Coalfield, Lancashire and Cheshire Historical Society*, vol. ii, 1861-2, passim.
31 L.M.F. conference, 19 January, 1889; S. Woods' annual report, 1890.
32 T. Ashton, op. cit., vol. i, p. 122.
33 Ibid., vol. i, p. 166.
34 T. Ashton, ibid., vol. i, p. 158.
35 *Workmen's Times*, 17 September, 1892.
36 Thomas Ashton, op. cit., vol. i, p. 66.
37 L.M.F. Conference, 6 June, 1894.
38 L.M.F. Conference, 11 March, 1893.
39 L.M.F. Conference, 6 October, 1893.
40 L.M.F. Conference, 4 January, 1895.

NOTES ON CHAPTER THIRTEEN

1 *Wigan Observer*, 29 January, 1898.

2 B. McCormick and J. E. Williams, Miners and the Eight-Hour Day, *Economic History Review*, 1959, p. 233.

3 T. Ashton, *Three Big Strikes in the Coal Industry*, vol. ii, p. 248, reported that weighing was permitted at Burnley collieries only after June 1902. In other words, forty-two years after the Act which authorized the appointment of checkweighmen.

4 These loopholes were to some extent plugged by the Mines Act 1894, which made it illegal for an owner or manager to exercise 'improper influence' on the election of a checkweighman.

5 Roy Gregory, *The Miners and British Politics, 1906-1914*, p. 12.

6 A M.F.G.B., deputation to meet Lord Salisbury, 14 March, 1895, quoted by T. Ashton, op. cit., vol. ii, p. 46.

7 Henry Pelling, *Social Geography of British Elections, 1885-1910*, p. 266.

8 W. F. Monypenny and F. E. Buckle, *The Life of Benjamin Disraeli, Earl of Beaconsfield*, vol. ii, p. 712. I made this quote because I think it represents a view widely held among miners. I have not been able to trace the source of Alexander MacDonald and am inclined to agree with Mr. Trevor Lloyd, *The General Election of 1880*, p. 47, that this was out of character. MacDonald appears to have remained true to his Liberal principles.

9 *Workmen's Times*, 2 September, 1893.

10 John Vincent, Second Act in Lancashire, *The Historical Journal*, vol. 11 1968, p. 43.

11 H. Pelling, op. cit., p. 241.

12 A. Redford, *Labour Migration in England, 1800-1850*, p. 113, noted that even in the first half of the nineteenth century Irish immigration in Lancashire tended to depress wage rates.

13 H. Pelling, op. cit., p. 266. It is interesting to note that Michael Davitt helped Labour candidates in Lancashire to secure the Irish immigrant vote in the 1906 general election. (See F. Sheehy-Skeffington's *Michael Davitt*, pp. 195-6.) The 1906 election was the first election in which they made significant gains in Lancashire.

14 Hyndman's letter to the *Pall Mall Gazette*, 28 January, 1881, cited by Tsuzuki, H. M. Hyndman and British Socialism, p. 37.

15 Hyndman to Marx, 15 January, 1881, *Karl Marx Correspondence*.

16 *Justice*, 19 September, 1885.

17 Ibid., 15 October, 1887.

18 Dona Torr, *Tom Mann and his times*, vol. i, p. 239. She is, of course, wrong when she contends that South Lancashire '*always remained* the strongest provincial centre of the S.D.F.'. As I show in the next chapter, this distinction later belonged to North Lancashire.

19 Dona Torr, op. cit., vol. i, p. 251.

20 *Justice*, 12 May, 1888.

21 *Workman's Times*, 1 April, 1893.

22 Walter Kendall, *The Revolutionary Movement in Britain 1900-1921*, p. 265.

23 *Workmen's Times*, 2 September, 1893.

24 *Labour Leader*, 8 June, 1895.

25 *Workmen's Times*, 9 September, 1893.

26 *Labour Leader*, 27 October, 1893.

27 *Workmen's Times*, 11 November, 1893.

28 *The Clarion*, 27 July, 1895.

29 C. Tsuzuke, op. cit., p. 97 and p. 284. He is quoting from the *Burnley Socialist*, 7 October, 1893, and the *S.D.F. annual report, 1894*, p. 16. On 11 November, 1893, the *Burnley Socialist*, which had a circulation of three thousand copies, claimed that there were nearly one thousand members in the Burnley branch alone.

30 *Workmen's Times*, 2 September, 1893.

31 T. Ashton's secretary's statement, May 1897. On one occasion Ashton did openly declare his Tory principles. In 1883, he wrote supporting T. H. Sidebottom, the Conservative candidate for Stalybridge and Duckinfield. His letter, published as a leaflet with the title 'Another underhand radical dodge', was written on official union notepaper with the heading Lancashire & Cheshire Miners' Federation, Ashton-under-Lyne District, 179, Grey Mare Lane, Bradford.

Ashton's conduct aroused criticism since it appeared to suggest that the union was officially supporting the Conservative candidate. McKay. president of the Ashton-under-Lyne Miners' Association, wrote an open letter to William Sumners, the Liberal candidate, saying, 'I think that Mr. Ashton has acted very unfairly.'

32 Minutes of Ashton and Haydock district union meeting, 29 March, 1890.

33 Minutes of Parliamentary Committee, Ashton and Haydock district union, 1 May, 1890.

34 Lancashire Miners' Federation conference, 24 May, 1890.

35 *Labour Tribune*, 24 May, 1890.

36 Ibid., 26 July, 1890.

37 Ibid., 2 August, 1890.

38 Ibid., 16 August, 1890.

39 Ibid., 9 August, 1890.

40 *Wigan Comet*, 23 July, 1892.

41 *Manchester Courier*, 1 April, 1881.

42 *Manchester Examiner*, 1 April, 1881.

43 See H. L. P. Hulbert, *Sir, F. S. Powell, Bart., and M.P.: A Memoir by his Nephew*. Also, Dennis Brown, *The Labour Movement in Wigan 1874-1967* (M.A. thesis, Liverpool University, 1969), pp. 29-32.

44 *Wigan Comet*, 23 July, 1892.

45 Ibid.

46 *Wigan Examiner*, 17 September, 1892.

47 *Wigan Observer*, 21 January, 1893.

48 *Wigan Comet*, 11 March, 1893.

49 *Labour Tribune*, 7 January 1893. It is interesting to note that some Conservative leaders favoured the notion of some sort of Conservative Labour organization, although it seems that they envisaged its activities in electoral rather than industrial terms. In their book, *Angels in Marble*, Robert McKenzie and Allan Silver quote a letter from Lord Salisbury to A. A.

Douglas saying, 'As far as I am concerned, my vote is entirely in favour of Conservative Labour candidates.' (p. 40.)

McKenzie and Silver point out that the idea was dropped when trade unions affiliated to the Labour Party. (p. 42.) It seems that only in Lancashire did the proposal gain any solid support.

50 *Wigan Comet*, 21 January, 1893.

51 *Workmen's Times*, 6 February, 1892.

52 *Workmen's Times*, 9 April, 1892.

53 *Burnley Socialist*, 1 December, 1893.

54 Lancashire Miners' Federation leaflet, published by Thomas Ashton, 12 April, 1893.

55 R. Page Arnot, *The Miners*, vol. i, p. 237.

56 Ibid., p. 238.

57 *The Labour Elector*, November 1893.

58 Ibid.

59 *Workmen's Times*, 24 February, 1894.

60 *Labour Leader*, 23 March, 1895.

61 *The Clarion*, 26 November, 1898.

62 I.L.P. annual report, 1897, p. 15-16. Keir Hardie's comment on Sam Woods return as a Lib.-Lab. M.P. for Walthamstow.

63 Beatrice Webb's Diaries, 15 January, 1896.

64 *Wigan Comet*, 1 October, 1892.

65 *Wigan Observer*, 9 July, 1892.

66 Ibid.

67 Ibid., 30 January, 1895.

68 *Sunday Chronicle*, 26 January, 1895.

69 West Lancashire miners' delegate meeting, 12 January, 1895.

NOTES ON CHAPTER FOURTEEN

1 Letters Campbell-Bannerman to Gladstone, 19 October, 1901, and Gladstone to the secretary of the Dewsbury Liberal Association, 3 October, 1901, mentioned in F. Bealey and H. Pelling, *Labour and Politics, 1900-1906*, p. 132.

2 *Pendlebury Pioneer*, December 1901.

3 R. C. K. Ensor, *England, 1870-1914*, p. 222.

4 Report of the Conference on Labour Representation, 27 February, 1900, p. 21.

5 Report of the 4th annual conference of the Labour Representation Committee, 4 February, 1904, p. 6.

6 For full accounts of the arrangements reached between the Labour Representation Committee and the Liberal Party see Philip P. Poirier, *The Advent of the Labour Party*, chapter X, and F. Bealey and H. Pelling, op. cit., chapter VI.

7 Herbert Memorandum of interview with MacDonald on 6 September, 1903. H. G., vol. cxvi, cited Philip P. Poirier, op. cit., p. 252.

8 Philip P. Poirier, op. cit., p. 252.

9 Ibid.

10 Labour Representation Committee Minutes, 2 December, 1904.
11 F. Bealey and H. Pelling, op. cit., p. 246.
12 T. C. Barker and J. R. Harris, *A Merseyside Town in the Industrial Revolution—St. Helens, 1750-1900*, p. 475.
13 Roy Gregory, *The Miners and British Politics, 1906-1914*, p. 82.
14 *Daily News*, quoted *Wigan Observer*, 2 March, 1913.
15 W. Kendall, *The Revolutionary Movement in Britain, 1900-1921*, chapter 1: Stress and Strain in the S.D.F.
16 Eric Hobsbawm, *Labouring Men*, p. 232.
17 Two articles in the *New Review*, quoted *Wigan Comet*, 5 October, 1889.
18 *The Times*, 5 January, 1898.
19 J. M. Davidson, *Mining Rents and Royalties*, p. 1.
20 English Land Restoration League, 4th Report, 1886-7, cited in E. Eldon Barry, *Nationalisation in British Politics—the Historical background*, Jonathan Cape, London (1965) p. 126.
21 R.C. on Mineral Rights and Royalties, 1890-1, s. 8659.
22 Ibid. Q.8510.
23 Ibid. Q.8517.
24 Ibid. Q.8565-6.
25 For the most detailed account of the development of miners' ideas on this subject see E. Eldon Barry, op. cit., chapter iv, *Nationalisation of Coal-mines*. The first advocacy of coal nationalization I have come across was made by Keir Hardie in his journal, *The Miner*, July 1887.
26 M.F.G.B. Annual Conference Report, 1894; *The Times*, 20 January, 1894.
27 Lancashire Miners' Federation minutes, 16 November, 1895.
28 *Wigan Comet*, 24 December, 1892.
29 T. Ashton, *Three Great Strikes in the Coal Industry*, vol. ii, p. 81. The Scottish miners had a resolution very similar to Lancashire's before the conference.
30 For a full discussion of deference voting see Eric Nordlinger, *The Working-Class Tories*. Also Robert McKenzie and Allan Silver, *Angels in Marble*.
31 *Daily Mail*, 13 December, 1909.
32 L. Thompson, *Robert Blatchford*, p. 27.
33 James Roll, *Second International*, p. 123.
34 *Morning Post*, 6 July, 1910.
35 *Workmen's Times*, 18 June, 1892.

NOTES ON CHAPTER FIFTEEN

1 *The Edinburgh Review*, January 1860, pp. 7-8.
2 Peter Mathias, *The First Industrial Nation*, p. 135.
3 Eric Hobsbawm, *Industry and Empire*, p. 40.
4 Figure quoted by E. C. Midwinter, *Victorian Social Reform*, Longmans, London (1969) p. 10.
5 The three investigations were: Report of the Select Committee of the

Health of Towns (1840), Report of the Sanitary Condition of the Labouring Population (1842) and Inquiring into the State of Large Towns and Populous District (1844-5).

6 Frederick Engels, *The Condition of the Working Class in England in 1844*, Allen & Unwin, London (1948 ed.) p. 146.

7 Matthew Arnold, 'The Future of Liberalism', an essay in his *Irish Essays*, p. 111.

8 John Ruskin, *Fors Clavigera*, vol. i, p. 270. William Dodd's book, *Factory System*, published in 1842, gives a very similar description of Wigan: 'These streets were unsewered, unpaved; every few yards a pool of stagnant water, and heaps of accumulated filth of every description.'

9 *Wigan Observer*, 15 May, 1875. The Board of Health regularly issued figures for the death rate in fifty towns in England, and Wigan frequently either topped the list or came near the top. It is interesting to note that the mortality figure given in here—42·1 per thousand—is significantly above the death rate of 22·8 per thousand for the Wigan registration district in 1875. (Cited *Journal of the Royal Statistical Society*, March 1880.) This discrepancy can probably be attributed to the fact that the Wigan registration district included many rural areas that were very much healthier than the town itself.

10 Ibid., 9 May, 1873.

11 *The Lancet*, quoted in the *Wigan Observer*, 21 February, 1874.

12 Medical Officer of Health's report on Upholland, 11 August, 1874.

13 *Wigan Observer*, 11 July, 1872.

14 Ibid., 24 October, 1874.

15 Quoted in P. E. H. Hair, *The Social History of the British Coalminers, 1800-1850*. (Ph.D. thesis, Oxford.)

16 *Wigan Observer*, 30 May, 1872.

17 *Liverpool Weekly Albion*, 2 August, 1873.

18 T. C. Barker and J. R. Harris, *A Merseyside Town in the Industrial Revolution*, p. 278.

19 *Manchester Guardian*, 29 August, 1874.

20 *Wigan Observer*, 19 April, 1873. A more detailed study of the subject of kicking has been written by Arthur J. Hawkes, 'Purring or Up-and-down Fighting in Lancashire', published in the *Wigan Examiner*, 20 March, 1943.

21 *Wigan Observer*, 19 April, 1873. Professor Perkin has pointed out to me that duelling with intent to kill or grieviously hurt no longer took place in upper-class English circles by the Eighteen-Seventies. Presumably, however, serious injury or death could be the unintentional outcome of, say, sword-fencing—and definitely was the outcome of many 'purring' matches.

By an Act of 1882, a man found guilty of kicking could be sentenced to transportation for life or three years' imprisonment with hard labour. Frank Ormerod, in his *Lancashire Life and Character* said: 'Many a victorious combatant has had to flee the country after one of these contests to escape arrest for manslaughter, or lie in hiding for weeks and months on the moors until the law had relaxed its efforts.'

22 E. C. Midwinter, *Law and Order in Early Victorian Lancashire*, p. 26.

23 Ibid., p. 28. Mr Midwinter states, 'Thomas Latham, the new Chief Constable, did little to abate the criminal tendencies of his troop.'

24 *Wigan Observer*, 12 August, 1854.

25 Quoted in Professor Harold Perkin, *The Origins of Modern English Society, 1780-1880*, p. 171.

26 E. C. Midwinter, *Victorian Social Reform*, p. 10.

27 T. C. Barker and J. R. Harris, op. cit., p. 270.

28 Peter Higson, Mines Inspector's Report for 1870, p. 5.

29 Rev. T. R. Fergie's evidence to the Royal Commission on the Labour Laws, 30 June, 1875, Manchester.

30 The Amalgamated Association of Miners even passed resolutions at its conferences calling for town improvements and more houses.

31 *Miners' Advocate*, October 1845.

32 The figures were given in the Report of the Mines Inspectors for 1845, p. 5. It should be remembered that these statistics only relate to underground labour. P. E. H. Hair's estimates in his thesis that there were only about 150,000 surface workers at collieries in 1841. In Lancashire the number of women who worked as surface workers seems to have remained fairly constant after 1842. Since the population increased, this meant a smaller proportion of female labour was employed at collieries as the century continued.

33 Children Employment Commission, 1842, vol. i, p. 105.

34 Dr J. D. Marshall, The Lancashire Rural Labourer, *Transactions of the Lancashire and Cheshire Antiquarian Society*, 1961, p. 108.

35 The Inspector of Mines Report, 1846, p. 5.

36 J. Dickinson's Mines Inspector's Report, 1846, p. 5.

37 J. H. M. Bankes, *The Nineteenth Century Colliery Railway*, Trans. of the Lancashire and Cheshire Historical Society, vol. 114, p. 174.

38 The Inspector of Mines Report, 1846, p. 50.

39 Children Employment Commission, 1842, vol. i, p. 285.

40 *Manchester Courier*, 16 July, 1875.

41 *Labour Tribune*, 24 May, 1890.

42 T. Ashton, *Three Big Strikes in the Coal Industry*, vol. i, p. 105.

43 Figures quoted in the *Wigan Observer*, 28 December, 1963, from a Ministry of Education table which gave expenditure per 1,000 children of school age in 1962-3. Since then, the situation has somewhat improved. The Institute of Municipal Treasurers and Accountants have issued the Education Statistics for 1967-68. These show that, of the eighty-two county boroughs, Wigan is now eightieth in its expenditure for a thousand pupils in primary school. St Helens has risen even further up the list.

44 Labour's Social Strategy, published in August 1969. The indices of inadequacy and deprivation were compiled by Bleddyn Davies. They include eighteen items such as higher than national average mortality figure, below average income, the proportion of sub-standard housing, the incidence of T.B., proportion of children staying at school after the age of sixteen, etc.

NOTES ON CHAPTER SIXTEEN

1 H. S. Jevons, *The British Coal Trade*, p. 73, also Walcot Gibson *Coal in Great Britain*, p. 248.

2 Royal Commission on Mining Royalties, 1890, Appendix A, xi. The Testimony of John Knowles of Pendlebury.

3 Ibid., Appendix B. Evidence of the South Lancashire and Cheshire Coal

Association. (In fact, the Association included owners from West Lancashire and North-East Lancashire.)

4 Ibid., Appendix B.

5 H. S. Jevons, op. cit., p. 73.

6 R.C. on Mining Royalties, 1890, Q.2801. Testimony of Alfred Hewlett.

7 *Daily News*, quoted *Wigan Observer*, 2 March, 1913. Big coalowners were reluctant to recognize trade unions. Pilkington Brothers only conceded recognition in 1917. See T. C. Barker's *Pilkington Brothers and the Glass Industry*, p. 201.

8 Celia Batty in *A Study of Coal-Mining in the North-east District of Lancashire* does give a reasonably full account of mining dynasties from Elizabethan times.

9 Account of Extraordinary meeting of Wigan Coal & Iron Company, 13 January, 1870. The pamphlet is contained in the Wigan Reference Library's Documents relating to Mining and Manufacture, vol. 10.

10 *Wigan Comet*, 2 May, 1893.

11 Ibid., 1 October, 1892.

12 Ibid.

13 L.M.F. conference, 28 May, 1898.

14 L.M.F. conference, 23 July, 1898. Communication from Platt Bridge. Also, around this time there were numerous cases of aged and infirm colliers who had been dismissed from their work, applying to the Federation for relief, and these are, of course, contained in the minutes of the L.M.F.

15 Minutes of Executive meeting, 28 May, 1898.

16 Minutes of L.M.F. Executive Meeting, 16 April, 1898.

17 Sam Woods' pamphlet was mentioned at the L.M.F. conference, 22 January, 1898.

18 L.M.F. Executive meeting, 29 January, 1898.

19 The strike of surgeons, unwilling to collaborate with the union-sponsored relief scheme, was reported in *The Clarion* of 12 August, and 26 August, 1898. Stephen Walsh said that the surgeons did not want more money but to sabotage the scheme.

20 Henry Cunninghame's reply to Philip Stanhope, M.P. in the House of Commons, 2 September, 1897. Also, Home Secretary's letter to Thomas Ashton, 22 September, 1897: 'I am not even satisfied that a majority of the miners affected by the exemption desire the change.'

21 L.M.F. Executive meeting, 13 May, 1898. It should be noted that the Clifton mentioned was a colliery near Burnley, not the township in the south of the county.

22 L.M.F. Executive meeting, 19 February, 1898.

23 L.M.F. conference, 20 August, 1898.

24 R. Handsley's letter to Thomas Ashton, 20 June, 1898.

25 L.M.F. conference, 20 August, 1898.

26 Thomas Ashton, *Three Great Strikes in the Coal Industry*, vol. ii, p. 137.

27 Thomas Ashton, ibid., vol., ii, p. 248-9.

28 John Leeming had come in for attack from socialists for his Liberal views as long ago as 1893: See *The Burnley Socialist*, 14 October, 1893.

29 *The Clarion*, 25 June, 1898, and in most issues in July, the advantages of the scheme were outlined.

30 L.M.F. conference, 30 April, 1898.

31 Ibid. Quoted in P. J. King's pamphlet, *Good and Bad Federation*, p. 11.

32 P. J. King, ibid., p. 12.

33 *The Clarion*, 30 July, 1898.

34 Ibid., 23 July, 1898.

35 William Hallam, *Miners Leaders*, p. 50.

36 R. Page Arnot, *The Miners*, vol. i, p. 63.

INDEX